STILL USEFUL in an AI WORLD

The Human Playbook for the Age of AGI

Ken Konet, M.Ed., MBA

Humbolton Press
April 2026

STILL USEFUL in an AI WORLD

The Human Playbook for the Age of AGI

Humbolton Press

www.humbolton.com

First Edition: April 2026

ISBN (Paperback): 978-1-966703-32-7

Disclaimer.

This book contains the author's opinions, research, and ideas on career transition, workplace automation, and economic trends. It is intended for informational and educational purposes only. The author is not a licensed financial advisor, attorney, tax professional, or career counselor, and nothing in this book should be taken as professional financial, legal, medical, or career advice. Every reader's situation is different. Readers should consult qualified professionals before making significant career, financial, tax, or legal decisions. The author and publisher disclaim any liability, loss, or risk incurred as a consequence, directly or indirectly, of the use or application of any information contained in this book.

The five recurring case studies presented in this book (Dave Palmieri, Marisol Quintero, Jamal Bankole, Dr. Eleanor Voss, and

Kevin Huang) are composite portraits drawn from the author's consulting work, interviews, and research. Names, locations, employers, and identifying details have been changed or invented to protect individual privacy. Any resemblance to specific living persons is coincidental.

All trademarks, product names, company names, and brand references mentioned in this book are the property of their respective owners and are used for identification purposes only. Their use does not imply endorsement.

Cover and Interior design: Humbolton Press
Printed in the United States of America.

Abstract

Still Useful in an AI World: The Human Playbook for the Age of AGI is a career survival guide for knowledge workers navigating the most consequential labor transition in a century. Drawing on current economic research, peer-reviewed studies on generative AI and labor, and real-world case studies, corporate instructional designer Ken Konet argues that the artificial intelligence revolution is not a distant threat but a present-tense restructuring of white-collar work, already executing on quarterly timelines across every knowledge-based industry in the United States.

Rather than offering generic reassurance, the book identifies three durable human capacities that continue to command an economic premium as automation expands: Physicality (embodied work in unpredictable environments), Accountability (regulated judgment that requires a human signature and carries personal liability), and Taste (curatorial and editorial discernment at a scale where machines can produce infinite mediocrity).

Across twenty-two chapters built around the author's instructional design spine of hook, concept, example, application, reflection, and action, readers develop diagnostic clarity about their own exposure, build practical AI literacy (including prompt architecture and multi-agent workflow construction), and execute a structured pivot toward roles that will remain economically durable through 2040 and beyond. Five recurring case studies (a mid-career insurance underwriter, a plumbing business owner, a solo marketing operator, a senior psychiatrist, and a recent college graduate) trace the specific decisions that separate successful pivots from passive drift.

The book includes eight appendices of usable tools: a 90-Day Human Hardening Plan, an Industry Survival Matrix across sixteen sectors, a Prompt Architecture Starter Kit, a reading list of primary sources, a glossary, a 25-Question Human Skills Self-Assessment, pivot and credentialing resources, and the author's personal playbook. Approximately 62,000 words. Written for mid-career professionals in exposed industries, early-career workers in a restructuring labor market, and anyone planning a household's economic future through the AGI transition.

PART ONE

The AGI Shock

The robbery is already happening. The first job is admitting it.

Table of Contents

Chapter 1: *Welcome to the Heist*

The robbery has already happened. You are in the middle of it. Some of your coworkers are getting pickpocketed in slow motion — right now, today, during this meeting you are half-paying-attention to — and they have not looked down yet.

Here is how it goes.

Brad from underwriting has been at the firm for nineteen years. He knows the policies, the regional rules, the weird carve-outs for commercial trucking fleets in three specific states. He knows which adjusters will call you back and which ones will ghost you until Thursday. Brad is valuable. Brad is safe. Brad has a kid starting Ohio State next fall.

Brad is also, unbeknownst to Brad, already a cost center with an expiration date.

Last quarter, the firm licensed a system. It does not have a name because that is not how software licenses work at mid-market insurance firms. It has a line item and a SKU and a three-year contract. The system, over the next eighteen months, is going to absorb roughly 73% of Brad's weekly workflow. Brad's manager does not know this yet in those terms, because Brad's manager is also busy getting robbed, just one rung up the ladder.

Nobody is going to walk into Brad's cubicle with a gun. Nobody is going to fire him on a Tuesday. Brad is going to notice, six months from now, that his email volume has dropped. That his one-on-ones have gotten shorter. That Julie from HR keeps bringing up "career pathways." By the time Brad figures out what is happening, the robbery will be 90% complete, and Brad will be standing in his kitchen at 2 a.m. Googling "jobs for insurance professionals over 45."

Welcome to the heist.

— — —

A Brief Word Before We Continue

Yes. I know. You picked up this book because you wanted someone to tell you what to do, not to scare the hell out of you in the first three paragraphs. Fair.

Here is the deal. I am not going to lie to you for three hundred pages. Not about this. You are smart enough to know when someone is soft-pedaling, and I respect you too much to waste either of our time. So we are starting with the truth, which is that the biggest economic transition of your working life is happening to you right now, whether or not you have noticed, and the first rule of surviving it is admitting that it is happening.

Once we have done that — and it will take us about two thousand more words — we will get to the plan. I promise there is a plan. I have one for the next two hundred and fifty pages. So relax. Grab a coffee.

— — —

The Heist, Properly Framed

A heist works because of three things: misdirection, specialization, and speed.

Misdirection is what we talk about when we talk about AI. For three years now the conversation has been dominated by Drake deepfakes, chatbots that hallucinate case law, and earnest LinkedIn posts from guys with ring lights explaining why "AI won't replace you — a human using AI will." That last one, by the way, is true for about 17% of knowledge workers and catastrophically false for the rest. We will get there.

While you were watching the chatbots write bad sonnets, the actual robbery has been happening in the back office. It is happening inside Excel, inside Salesforce, inside legal discovery platforms, inside medical coding software, inside the guts of every SaaS tool your company pays seven figures for. It is happening because the companies that make those tools have been quietly integrating large language models and agent workflows into their product stacks since roughly Q3 2023, and those integrations are now, in 2026, mature enough to eat real headcount.

Specialization is the part that gets overlooked. AI is not coming for "your job" as some amorphous blob. It is coming for specific tasks inside your job, in a specific order, starting with the ones that are most repetitive, most describable, and most boring. And here is the ugly secret: a lot of us built careers out of exactly those tasks. That is what we got hired for. That is what we got good at. That is what the firm paid for. We are not being attacked as whole humans. We are being atomized into tasks, and the tasks are being sorted into "keep" and "automate" piles. Right now. While you read this.

Speed is the third thing, and it is the thing nobody wants to talk about. The Industrial Revolution took about eighty years to fully redistribute labor from farms to factories. The personal-computer revolution took thirty. The internet took fifteen. Generative AI and agent systems are currently operating on a timeline measured in quarters. Not decades. Not even years. Quarters.

So, we are being robbed, by a thing we are not looking at, in a way we cannot see, at a speed we are not built to track.

If that sounds grim, hold on. This is just the diagnosis. The cure starts around page sixty.

— — —

Is This Just Hype?

I hear you. I do. And I was asking the same question in 2022, which means I get to rub it in now.

Here is what is not hype, and what you should look at if you still think this is all going to blow over.

The Stanford AI Index. Every year. Look at the benchmarks. Look at the rate of change. In 2021, the best AI system in the world scored below human performance on nearly every cognitive benchmark Stanford tracks. By 2024 it was at or above human performance on most of them. By 2026, the benchmarks themselves had to be redesigned because the AI kept saturating them. This did not happen on a Reddit thread. It happened in peer-reviewed work you could verify right now in four seconds if your boss would stop sending meeting invites.

The Anthropic Economic Index. This is my favorite data set for a very specific reason: it uses actual, real-world AI usage data from millions of users to map which occupations are using AI most heavily and for what tasks. The top of the list is predictable — software engineers, marketers, writers, analysts. The middle of the list is where your jaw drops. Legal services. Financial services. Management consulting. Pharmacy. Customer-service management. Insurance underwriting. (Sorry, Brad.) Not the stuff nobody saw coming. The stuff people kept insisting was safe.

Goldman Sachs, 2023: roughly 300 million full-time-equivalent jobs globally exposed to automation by generative AI. Not eliminated — exposed. Meaning significantly reshaped, partially automated, or made vulnerable to displacement. They did not pull that number out of a hat. They went through the Occupational Information Network task list, rated every task for AI exposure, and multiplied. Their methodology is public. You can argue with the assumptions. You cannot argue with the math.

MIT, Harvard, Princeton, and a dozen others, 2023–2025: consistent productivity deltas of 20–40% on white-collar knowledge work with AI assistance. Some studies as high as 60%. Let me translate that out of academia. If three people on a team can suddenly do the work of five, guess what happens to the other two.

And finally, the single most chilling data point, which is buried in half these reports and which nobody wants to talk about: *the productivity gains are larger for less-experienced workers*. Meaning the AI is eating the bottom of the career ladder. The junior roles. The ones you used to take to learn the craft before you became the senior person. Those rungs are disappearing in real time.

So is it hype?

Some of it. The "AI will solve cancer by Thursday" takes are hype. The "AI will replace 50% of all jobs by 2027" takes are hype. The "AGI is two weeks away" takes, depending on which Twitter account you read, are hype.

But the boring, grinding, task-by-task automation of everything that currently pays your mortgage? That is not hype. That is a balance-sheet line item at every S&P 500 company, and it is executing on schedule.

— — —

The Three Things That Stay Human (For Now)

You picked up a survival manual, so here is the first shape of the answer. I am not going to fully explain any of these yet. We are going to spend the next twenty chapters earning them. But I want you to have them in your pocket before we get to the bad parts.

One: Physicality. You are made of meat, in an unpredictable physical world. Robots are getting better every year, but the economics of replacing a human body in a messy space, at scale, for less than the cost of the human, are brutal. We will call this the Physicality Wall. It is the reason a plumber in Phoenix is going to do just fine while the MBA in Phoenix is updating her LinkedIn.

Two: Accountability. You can be sued. You can be fired. You can be shamed. You can be thanked. You can sign a document and be held legally responsible for it. An LLM cannot. Not one of those. And every regulated industry on Earth — law, medicine, finance, engineering, construction, transportation — requires a human signature at the bottom of the page, and that requirement is getting *stronger*, not weaker. The signer is the premium.

Three: Taste. When AI can produce ten thousand logos, ten thousand emails, ten thousand pitch decks, ten thousand mediocre screenplays in a weekend, the bottleneck shifts from production to judgment. Knowing *which one is good*, and *what to make in the first place*, becomes the scarcest skill. Critics outlive creators. Editors outlive writers. Curators outlive collectors. This is where the money goes.

Physicality. Accountability. Taste. Write those three down. We are coming back to them.

— — —

This Book Is Your Playbook

I am an instructional designer by trade, which is a fancy way of saying I get paid to figure out how to teach adults things they really do not want to learn but desperately need to know. I have spent twenty-five years structuring information so that working professionals actually absorb it, and I have built this book around the same spine I build every corporate training program around. Hook, concept, example, application, reflection, action. Every chapter ends with a Monday-morning move. Every chapter has a diagnostic. Every chapter names one thing you can stop doing and one thing you should start doing, today, without permission from anyone.

You are going to meet five people over the course of this book.

Dave, the insurance guy we opened with. Marisol, a plumber in Phoenix who is building a small empire and has never once thought about AGI except to laugh at the people panicking about it. Jamal, a twenty-nine-year-old marketing founder who runs a four-million-dollar business with two human employees and forty AI agents. Dr. Eleanor Voss, a sixty-three-year-old psychiatrist in Boston who has a waitlist of eighteen months and no intention of retiring. And Kevin, a college senior who just figured out that his entire business degree is about as valuable as a Blockbuster card. (Look it up. Make your kid explain it to you.)

Those five are going to come with us through the book. Their situations map to most readers' situations. Their pivots are real. Their lives are the before-and-after of the three pillars.

— — —

The Diagnostic (First Pass)

Answer the following in the margin of this page. Or, if you are on a Kindle, in the notes. Or just in your head while you pretend to listen in a three o'clock meeting.

- Does your job involve mostly moving information from one screen to another?
- Could someone read a flowchart of your weekly tasks and fully understand your role?
- When you describe what you do to your parents, do you use phrases like "I handle reports," "I coordinate," or "I manage the process"?
- Do you ever physically touch the thing your work produces?
- Is your final output something a human signs their name to, or something a system spits out on its own?

If you answered yes to the first three, you are in the blast radius. That is fine. Most of us are. Awareness is the first move.

If you answered yes to the last two, you have more runway than you think. We will sharpen that edge over the next twenty chapters.

— — —

Monday Morning

Before you go back to pretending to pay attention to your Slack notifications, here is what you do tomorrow morning.

Open a blank document. Name it "My Job, Honestly." Write down every recurring task in your week. Not the meta stuff — not "provide strategic leadership" or "drive alignment" — the actual tasks. Review this report. Send that email. Update this spreadsheet. Attend that meeting. Brief that person. Draft that memo.

Next to each task, write a letter: A, B, or C.

- **A** means: this task could be done by an AI tool currently available, in 2026, for under $100 a month.
- **B** means: this task requires me, but mostly because of institutional inertia, and it will be an A within eighteen months.
- **C** means: this task genuinely requires a human body, a human signature, or human judgment in a messy situation.

Do not lie. Nobody is going to see this list but you. Lying to yourself at this stage is the single most expensive thing you can do. The C tasks are your runway. The A and B tasks are the evacuation route.

— — —

Close

You were robbed today. You will be robbed tomorrow. You will be robbed again on Friday.

The good news about heists is that they have survivors. In fact, every heist movie ever written has a survivor, and usually a few of them, and they are usually the ones who saw the con coming before anyone else did.

That is you now. You saw it. Not everyone at your company will. Some of your colleagues will keep their heads down, keep executing tasks on the A list, and hope it all goes away. That is their decision, and I wish them well. You picked up this book, which means you have already made a different one.

You are not behind. You are early. The average American is going to figure this out about two years from now, which means right now, today, the world is quietly full of jobs, opportunities, and pivots that nobody is fighting over yet. You are reading a survival manual while most people are still watching the trailer for the movie.

Let us go.

Next chapter: how we got here, why knowledge work was always more fragile than it looked, and what the Industrial Revolution can teach us about what is about to happen to the MBA.

Chapter 2: *The Cortex Revolution*

The Industrial Revolution took your grandfather's back. This one is after your brain. Fair trade, I guess.

In 1811, a British weaver named Ned Ludd allegedly showed up at a factory with a hammer and broke some mechanical looms. The story is probably apocryphal, but nobody cares, because the story is too good. Ned became a meme two hundred years before memes were a thing. "Luddite" became the word we use for anyone who dares question the inevitability of the next great technology, which is ironic, because Ned Ludd was not anti-technology. He was pro-paycheck. He watched his trade, his apprenticeship, his way of life, and his small town get absorbed by machines that could do what his entire family did, cheaper, faster, and in factories where children lost fingers for twelve cents a day.

What people do not say about the Industrial Revolution is that it was a pretty bad time to be alive if you were a weaver. Real wages in Britain stagnated or dropped for roughly fifty years as machines absorbed textile work. Real wages. Stagnated. For half a century. Families broke. Kids worked in factories. London's life expectancy hovered around thirty-five. Dickens was writing true crime, not fiction.

Eventually — and this is the part people cite when they tell you "don't worry, history says we'll be fine" — wages rose, new jobs appeared, living standards climbed, and within a century the world was vastly richer than it had been.

Sure. And also, if you were Ned's kid, you were dead by forty.

"History says we'll be fine" is the kind of thing said by people who plan to be long dead before the actual adjustment is complete. You do not have that luxury. You have a mortgage, and the adjustment has to happen during your working life. So let us not get smug about the long arc of history. The long arc of history is great for your great-grandchildren. It is not great for you.

— — —

Muscle 1.0 vs. Cortex 2.0

Every previous industrial transition replaced some form of muscle.

The first replaced human muscle with mechanical muscle. Looms, steam engines, factory mills. If you were built to move things, lift things, weave things, mill things, you were in the crosshairs.

The second, roughly 1900 to 1950, replaced more muscle — horses, oxen, human labor — with internal combustion. Tractors killed the family farm. Cars killed the blacksmith. Refrigeration killed the iceman. (Yes, the iceman. That was an actual job. He delivered ice.)

The third, roughly 1970 to 2000, replaced *repetitive* cognition. Excel killed the bookkeeper. Word processing killed the secretarial pool. ERP systems killed the middle-tier clerk. You got the Microsoft Office suite, which — depending on how you count — eliminated somewhere between two and six million clerical jobs in the United States alone.

But notice the pattern. Every wave so far has still left plenty of room for one specific class of labor: the white-collar knowledge worker. The analyst. The manager. The professional. The person who takes information, thinks about it, and makes a judgment call about what to do with it.

That class — which includes roughly half of the American workforce and an even larger share of the American *wage bill* — has never before been in the direct blast radius of an industrial transition. The previous waves went *around* us, not *through* us. They chewed up other people's jobs. Our jobs got easier. Our salaries went up. Our degrees got more valuable.

The Cortex Revolution is coming through us.

This time the thing getting automated is not muscle. It is not even repetitive thought. It is *synthesis*. Judgment. Writing. Analysis. Pattern recognition. Problem-solving at the mid-to-high levels of cognitive complexity. The stuff you got your degree to do. The stuff you got promoted for. The stuff that, up until about three years ago, was considered fundamentally, essentially, quintessentially human work.

— — —

The Death of "MBA-Safe"

Let me introduce you to Dave Palmieri.

Dave is forty-seven, lives outside Cleveland, manages a claims team at a mid-market property and casualty insurance firm. Twenty-two years in the industry. Came up through the ranks, earned his CPCU designation, knows commercial trucking underwriting better than most people know their own children. Married, two kids, one at Ohio State and one at a branch campus. Mortgage. Honda Pilot. A reasonable retirement account that is not quite as large as the financial planner said it should be by forty-seven, but he is getting there.

For most of Dave's career, his job was considered safe. He had a specialty. He had a credential. He had seniority. Insurance is not sexy. Insurance does not get disrupted. Insurance is the definition of a boring, stable, American middle-class career, and Dave was a textbook case of doing everything right.

In Q2 of 2025, Dave's firm licensed a commercial underwriting AI platform. I will not name it because there are three of them and they all do roughly the same thing, but I will tell you what it does.

It reads applications. It cross-references them against forty data sources that used to take a junior underwriter three days to manually verify. It applies the firm's underwriting guidelines — guidelines that took Dave six years to fully memorize — consistently, without fatigue, at 3 a.m. on a Sunday. It produces a recommendation, a pricing, and a written justification. It flags exceptions to Dave or someone at his level.

When the platform launched, Dave's team had fourteen people. As of today it has nine. The firm's official position is that this is a "hiring freeze, not a reduction in force," which is the polite corporate way of saying they are letting the team shrink by attrition while the platform absorbs more of the workflow.

Dave has done the math. He is not stupid. He is an underwriter. Math is what he does. At the current rate his team will be four people in eighteen months. His own role, which is technically a manager of underwriters, is only necessary as long as there are underwriters to manage. Which there will not be.

Dave is the canary.

Not because he is special. Because he is so *normal.* Dave is every middle manager in every stable, boring, profitable industry in America. Dave is the insurance analog of your neighbor in finance, your brother-in-law in pharmaceuticals, your high-school friend in logistics, your cousin in HR. The careers that were supposed to be the safe ones. The careers our parents told us to get.

Those are the ones in the crosshairs now.

— — —

The Fragility We Did Not See

Here is the thing nobody in white-collar America wanted to admit until about twenty minutes ago: knowledge work was always more fragile than it looked. We just got away with it for forty years because the tools to replace us had not been built yet.

Think about it. What is most "knowledge work," really? It is moving information from one place to another with a thin veneer of judgment applied in between. Read this. Summarize it. Pass it up. Read that. Decide on it. Pass it down. Fill this form. Update that system. Attend this meeting. Brief that person. Memo. Memo. Memo.

For forty years we offshored huge chunks of this to lower-wage labor markets — India, the Philippines, Eastern Europe — and somehow still convinced ourselves that the core work was irreplaceable. The core work was never irreplaceable. The core work was expensive and hard to verify remotely. That is what kept it domestic. The minute somebody built a system that could do it reliably at American quality for pennies on the hour, the whole scaffold was going to come down.

That system got built. In 2023. It is called an LLM, and it is currently being integrated into every piece of enterprise software on Earth.

You know who was not surprised by this? Technologists who read Hans Moravec in 1988. Labor economists who read David Autor in 2003. McKinsey consultants who wrote reports in 2013 that nobody's managers read. Anyone who actually watched the trajectory of computational power and asked themselves the obvious question: what happens when we have enough compute to make a machine that does judgment work?

We got our answer. The machine costs twenty dollars a month and runs in a browser tab.

— — —

The Describable Trap

Here is the diagnostic I want you to internalize, because it is going to show up repeatedly in this book, and it is the single best heuristic for how exposed your job is.

If you can describe your job in a paragraph — if a smart twenty-two-year-old could read that paragraph, nod, and say "okay, I get it" — your job is highly automatable.

That sounds harsh. It is. But it is also incredibly useful, because it gives you a concrete yardstick to measure yourself against.

Jobs that are describable in a paragraph include most analyst roles, most coordinator roles, most junior-to-mid-level marketing roles, most entry-level legal work, most bookkeeping, most project management, most customer service, most HR administration, most sales operations, and most middle management. Yes, middle management. We will get to that.

Jobs that are *not* describable in a paragraph include most tradework, most direct patient care, most therapy and counseling, most high-level creative direction, and most senior executive roles where the work is fundamentally "making judgment calls in situations with incomplete information where you sign your name at the bottom and accept the consequences."

Notice the pattern.

The jobs that survive are the ones that are hard to describe because they involve a body, a signature, or a kind of judgment that is genuinely hard to compress into a paragraph.

The jobs that get eaten are the ones we thought were "thinking jobs" but were actually "information-moving-around jobs dressed up with a nice title."

— — —

The Diagnostic

Grab your job description. Not the one HR has on file. The real one — the way you would describe your work to a smart stranger at a wedding.

Ask yourself:

- Could a version of me, ten years younger, learn this job in under twelve months? If yes, AI can too. Probably already has.

- Does most of my day involve reading, summarizing, drafting, routing, reviewing, or approving? Yes? The AI is coming for those in that order.
- When I make a "judgment call," is it really a judgment call, or am I applying known rules to known situations? Be honest. You might be flattering yourself.
- Does my job involve a physical body, a regulated signature, or a relationship that has existed for longer than five years? If yes: more runway than you think.
- If I disappeared tomorrow, what exact task would nobody at the company be able to do? If the answer is "pretty much everything would eventually get figured out," you have a problem.

— — —

Monday Morning

Take your "My Job, Honestly" list from Chapter 1. Look at your A tasks — the ones an AI could do today for under a hundred dollars a month. For each A task, write down a single *adjacent* task that is *not* describable in a paragraph. Something that involves:

- Physical presence.
- A human relationship that took years to build.
- A signature on a decision with real consequences.
- A judgment call that no checklist could capture.

That adjacent task is your migration path. Your job is not "underwriter." Your job is "the person who owns the relationship with this specific client," or "the person who signs the final approval," or "the person who handles the exception cases that do not fit the model."

You need to start getting good at the adjacent thing. Today. Not when HR tells you to.

— — —

Science Anchor

If you want to chase this down, read:

- David Autor, *Why Are There Still So Many Jobs?* (2015). The labor economist's pre-GenAI view of what gets automated and what does not. It holds up remarkably well, with one massive caveat: he did not foresee how fast the judgment layer would fall.
- Daron Acemoglu and Pascual Restrepo on automation's displacement versus reinstatement effects. The punchline: displacement is winning, for the first time in economic history.
- The Anthropic Economic Index, which publishes real-world AI usage data by occupation. It is free. It is online. It will terrify you in a productive way.

— — —

Close

Dave closes his laptop at 5:47 p.m. on a Tuesday. His team had its weekly touchbase today. Nine people. Three months ago there were eleven. The two who left got pushed into early retirement with decent packages. It was fine. Everyone smiled. The office ordered Jersey Mike's.

He drives home in his Honda Pilot. The route takes him past the trade school on Highway 82, which just expanded its welding program because they cannot keep up with the demand. He does not notice. His mind is on his daughter's Ohio State tuition, and whether they should refinance, and whether his 401(k) is actually on track or whether the financial planner is just saying that to be nice.

Dave is not in trouble. Not yet. Dave has knowledge, experience, an industry designation, and enough runway to make a smart move if he starts now.

The problem is that nobody has told him he needs to make a move.

That is what this book is for.

Next chapter: why the plumber in Phoenix is laughing at all of us, and why a robot that can pass the bar exam still cannot fold your laundry.

Chapter 3: *Moravec's Paradox (or, Why the Plumber Is Winning)*

A robot can pass the bar exam. It cannot reliably fold your laundry. This is not a bug. This is the blueprint for your career.

Let me tell you about Hans Moravec.

Hans Moravec is a Slovenian-Austrian roboticist who spent most of his career at Carnegie Mellon building robots and, more importantly, writing about the weird thing he kept noticing while building them. In 1988 he published a book called *Mind Children*, which was half a survey of AI progress and half a philosophical treatise that people ignored for thirty-five years because it seemed like science fiction.

Inside it was a passing observation that, in hindsight, is probably the single most important insight in the history of labor economics. The short version: what is easy for computers is hard for humans, and what is easy for humans is impossibly hard for computers. A machine can do calculus. A machine cannot walk across a cluttered kitchen and reliably avoid a toddler.

Moravec's explanation was elegant and terrifying. Evolution spent five hundred million years optimizing animals for perception, locomotion, and physical manipulation of the world. Those processes run on massive parallel neural hardware that has been refined by life-or-death selection pressure for longer than there have been vertebrates. Evolution spent, generously, about fifty thousand years optimizing humans for symbolic reasoning, language, and logic. That hardware is bolted on top, kind of hacky, and honestly? Not that good at its job. We are impressed with ourselves for doing calculus because calculus is hard for us. But calculus is computationally trivial compared to the unconscious process of seeing a doorknob and turning it.

Which means — and here is the punchline — when we built artificial intelligence, we accidentally built machines that are really, really good at the things evolution did not have much time to optimize us for (the calculus, the chess, the legal briefs, the medical diagnostics) and really, really bad at the things evolution spent forever optimizing us for (the laundry, the doorknob, the cluttered kitchen, the one-year-old's grasp of the physical world).

This is Moravec's Paradox. And it is the single biggest determinant of which jobs survive the next twenty years.

— — —

The Physicality Wall

Okay. Buckle up. Sarcasm mode on.

For about a century now, American culture has treated "working with your hands" as a consolation prize. If you could not hack it in college, you went to trade school. If you were not cut out for the professional track, you went into the trades. Your mother cried a little at Thanksgiving. Your uncle, who was a union electrician making six figures, rolled his eyes and poured another drink.

Turns out Uncle Tony was right.

Because here is what is happening in 2026: the robots are coming for the lawyers, the accountants, the radiologists, and the junior consultants. The robots are *not* coming for the plumbers, the electricians, the HVAC techs, the welders, the elevator mechanics, the diesel mechanics, or the carpenters. Not because those jobs are less important. *Precisely because they are harder.*

That is not a typo. Bend a unionized pipefitter's job into a robot, and you will lose your engineering career trying. Bend a junior corporate attorney's job into a large language model, and a mid-tier law firm will pay you three hundred and fifty thousand dollars a year to do it.

This is the Physicality Wall. It has three layers, and they reinforce each other.

Layer one: unpredictability. A plumber walks into a house. Every house is different. Every crawl space is different. Every shutoff valve has been hacked together by three previous owners with four different tool philosophies. There is no training data for "this specific leaking pipe in this specific 1947 Cleveland bungalow with the weird electrical work the previous guy did in 1988." A robot needs a domain. A plumber needs a flashlight.

Layer two: dexterity. A state-of-the-art humanoid robot in 2026 has roughly the manual dexterity of a four-year-old with noise-canceling headphones on. It can do scripted motions in a factory. It cannot improvise a wire nut in the back of a ceiling fixture while holding a flashlight in its teeth. It probably will not be able to for another decade, maybe more. The hardware is not there. The software is not there. The training data is not there. The economics are not there.

Layer three: economics. This is the one nobody talks about. Even when robots *can* do physical work, the cost structure has to beat humans. A plumber in Phoenix makes eighty-five to a hundred and fifty dollars an hour fully loaded. A humanoid robot in 2026, if you can even get one, costs a hundred and fifty thousand dollars, requires an engineer to supervise it, and breaks in non-dramatic but expensive ways. The plumber is cheaper. The plumber is also more reliable, quieter, less litigious, and does not need a maintenance contract. The plumber wins.

For the Cortex Revolution to eat trade work, robots need to get dramatically cheaper AND dramatically better AND dramatically more reliable. That is three decades of improvement minimum, probably longer. For the Cortex Revolution to eat knowledge work, someone needs to type a prompt into a website. That happened in 2022.

— — —

Meet Marisol

Marisol Quintero is thirty-four years old and runs Quintero Plumbing in Phoenix. She has twelve employees and does residential and light commercial work. Her gross revenue last year was just north of two-point-one million dollars. She takes about three hundred and twenty thousand dollars in owner compensation. She drives a Ford F-250 that she bought used for cash. She owns a house.

She got her plumbing license in 2018 at the age of twenty-six. She spent two years apprenticing. She got her journeyman card in 2020 and her master's license in 2022. Total education debt: zero. Total time invested: roughly six years from starting apprentice to running her own shop.

For context, a junior associate at a major law firm in Phoenix in 2018 — Marisol's same starting year — graduated from law school with roughly two hundred and twenty thousand dollars in debt, spent three years in school, and started at about a hundred and ninety thousand dollars a year gross, minus aggressive debt service.

Eight years later, the lawyer is probably making around three hundred and fifty thousand dollars a year if she is still at the firm. The lawyer might still owe a hundred and fifty thousand on the loans. The lawyer is working sixty-hour weeks and is actively worried about whether her specialty — document review, contract negotiation, early-stage litigation research — is going to survive the GenAI transition.

Marisol is worried about the HVAC shortage in Arizona and whether she should expand into commercial refrigeration. She has no debt. She owns her house. She owns her business. She owns her trucks. Her biggest existential threat is not automation. It is finding enough qualified plumbers to hire, because *the entire trades industry in America is short by roughly five hundred and fifty thousand skilled workers as of 2026, and the shortage is widening every year.*

I am not telling you this to make you feel bad about your degree. I am telling you this because the cultural script you inherited is no longer matching the economic reality on the ground. The script said: go to college, get a white-collar job, never touch anything dirty, and you will be fine. The reality says: the dirty jobs are paying more, lasting longer, and enjoying better quality of life than most of the clean ones.

Marisol, for the record, laughs when I bring up AGI in conversation. Not mean laughing. Just the laugh of someone who got up at 6 a.m. to quote a job for flood damage repair in a house built in 1962 by a contractor who did not believe in building codes. "Yeah," she says, "let me know when the robot can crawl under that house." Fair.

— — —

The Hands-Plus-Judgment Audit

Here is this chapter's diagnostic.

For every task you do in a week, ask two questions:

1. Does this task require a physical body in a specific location at a specific time?
2. Does it require judgment that adapts to unpredictable circumstances in real time?

If your answer to both is yes, that task is deep in the moat. Robots will not touch it for at least a decade. Humans will be paid premiums for it, and rising premiums, for the foreseeable future.

If your answer is yes to one and no to the other, you are in a partial moat. Safe for now, vulnerable on a long horizon.

If your answer is no to both — if the task is entirely information-based, entirely describable, entirely predictable — you are in the desert. Not today's desert. Tomorrow's.

Now, this is the part that is going to sting: if you did the audit from Chapter 2, you probably already know what the results look like. Most white-collar jobs are mostly no/no tasks. A handful of yes/no or no/yes tasks. Very few yes/yes tasks.

The pivot, then, is not necessarily to become a plumber. Although if you are twenty-eight and reading this and considering law school,

I would very gently ask you to consider welding school instead. The pivot is to intentionally add yes/yes tasks to your work. Build in physical presence. Build in real-time adaptive judgment. Build in client-facing relationship work where your body in a room matters.

This is not vague corporate-speak. This is a concrete strategic move. The physical, the present, and the judgment-requiring are the only things the Cortex Revolution cannot currently touch.

— — —

What About Advanced Robotics?

I know. I can hear you. "But Ken, what about Boston Dynamics? What about Figure AI? What about Tesla's Optimus? What about Unitree?"

Yes. Good question.

The robotics industry is making real progress. Humanoid robots in 2026 can walk, navigate stairs, pick up specific objects, and execute scripted tasks in structured environments. Some of them are kind of impressive.

But here is the gap between "impressive demo" and "replaces a plumber."

- A plumber works in *unstructured* environments — houses and buildings built at different times, to different codes, by different people, with different quality levels. Each job is a one-off.
- A plumber uses *general-purpose* tools — wrenches, torches, snakes, meters — in combinations and adaptations that no robot has been trained for.

- A plumber deals with *unscheduled failures.* A robot that breaks down at 2 a.m. is a call to tech support. A plumber who breaks something at 2 a.m. is a plumber who fixes it and charges you double.
- A plumber has *legal accountability.* The guy who signs the inspection sticker is the guy on the hook if the house burns down. A robot cannot hold a license.
- A plumber *costs a hundred bucks an hour.* A humanoid robot currently costs six figures plus ongoing engineering support.

The gap between where robotics is in 2026 and where it would need to be to replace Marisol is, depending on which robotics researcher you ask, somewhere between ten and thirty years, and most of the researchers admit they have been wrong about every previous estimate, usually in the direction of "longer than we thought."

So no, a humanoid robot is not going to fold your laundry next Christmas. It is, possibly, going to fold the laundry at a warehouse with highly structured input conditions. That is a different thing.

If I had to pick a single bet for the next twenty years, I would bet that a kid who goes to welding school in 2026 will be making more money in 2040 than a kid who goes to a tier-two law school in 2026. That is not snark. That is a demographic, economic, and technological forecast with a lot of evidence behind it.

———

Monday Morning

List the three most physical-plus-unpredictable-plus-judgment-requiring activities your current job occasionally involves. They might be client visits. On-site troubleshooting. Showing up to the thing in person. Crisis meetings where someone has to make a call in the room.

Now go do more of those. This week. On purpose.

If your job does not have any of those, consider it your first real wake-up call of the book. You need to either (a) build them into your current role, (b) find a role where they are central, or (c) start thinking seriously about a pivot to a field where they exist.

— — —

Science Anchor

- Hans Moravec, *Mind Children* (1988). The original source. Still holds up. Short. Readable. Weird.
- MIT's annual robotics reports on humanoid platform progress.
- The Bureau of Labor Statistics occupational projections for skilled trades. Look at the ten-year growth estimates next to the ten-year growth estimates for, say, "paralegals." The gap is getting obscene.

— — —

Close

Marisol Quintero hangs up from a call, closes her laptop, pats her dog, and heads out to a job site at 4:30 p.m. to replace the water heater her crew has been struggling with. She will be home by 8. Her kids will have eaten pizza. Her spouse will have finished three loads of laundry. The laundry, as the robots have proven, gets done by humans.

She has not read a word of this book. She does not need to. She built a career in a domain that the Cortex Revolution cannot reach, and she did it in half the time and at a third of the cost of the white-collar track we all told her she should aim for.

If you have a kid in high school, consider her path before you write another tuition check.

That is not nostalgia. That is math.

Next chapter: why your degree, your skills, and your entire career trajectory are now running on a half-life measured in months — and why that is actually kind of good news if you know what to do with it.

Chapter 4: *The Six-Month Half-Life*

Remember when you picked a major? Cute.

In 1998, the average American worker changed jobs maybe three or four times in their career. The word "career" still meant something specific. You chose a major at nineteen, picked a profession at twenty-two, committed to a company or at least an industry at twenty-five, and by forty-five you were somebody's senior vice president with a corner office, a Rolodex, a Gantt chart, and a 401(k) that actually made sense.

In 2026, the average American worker under forty has changed jobs around six times already. The "career" concept is starting to feel like a museum exhibit — the kind of thing your grandparents describe to you while you nod politely and wonder how they ever tolerated fax machines. The only people still operating on the old career model are the ones who got into stable-enough positions twenty-plus years ago to ride it out, and even they are starting to notice that the pension fund is out of pensions and the new guy on the leadership team looks suspiciously like ChatGPT with a sport coat.

I am here to tell you something that is either liberating or terrifying, depending on your temperament.

Your skills have a half-life now. And that half-life is measured in months.

Let us break this down, because it is the central mechanical fact of working life in the Cortex Revolution, and getting your head around it changes everything about how you plan.

— — —

What "Half-Life" Actually Means (Without the Chemistry Lecture)

Half-life, in its original sense, is the time it takes for half of a radioactive substance to decay. Half your uranium is still uranium after, let us say, seven hundred million years. Half is something else.

For most of history, human skills had something that looked like a half-life, but it was measured in decades, if it existed at all. A blacksmith in 1800 could reasonably expect that the skills he learned at sixteen would still be the right skills at fifty-six. A mechanic in 1960 could reasonably expect that most of what he learned in his twenties would still be usable in his fifties, with some updates around the edges.

By 1990, the half-life had compressed to about ten years. By 2010, roughly five. By 2020, somewhere in the two-to-three-year range for most white-collar professions. By 2023, it was brutal. By 2025, the World Economic Forum was estimating that *40% of core job skills will change by 2030*, and honestly, that estimate feels conservative now that you can delete three years' worth of training content with a tool that did not exist last November.

Somewhere in 2025, for a lot of specific knowledge-work skills, the half-life dropped below twelve months.

For some — prompt engineering, specific automation workflows, particular coding patterns, certain financial modeling techniques — it is arguably under six.

Six-month half-life means half of what you know right now, today, that is directly useful to your job will be obsolete in six months. It does not mean you forget it. It means the world moved on. What replaced it is either automated entirely, or it is a completely new toolset, or the problem you used to solve has been redefined.

This is not hyperbole. This is not scare-tactic futurism. This is the actual lived experience of working knowledge workers in 2026. Ask any software engineer how their job has changed since January 2024. Ask any marketer. Ask any attorney doing anything that touches document review. They will tell you.

— — —

The Beta-State Career

Here is the mental shift that has to happen for you to survive this.

Stop thinking of your career as a *product*. Start thinking of it as a *continuously updated beta*.

Software companies figured this out twenty years ago. You do not ship a product anymore; you ship a service that is constantly being updated. Chrome is not "Chrome 1.0 (final)." Chrome is Chrome, and it is different every Tuesday. The idea that you would release software once and leave it alone is ludicrous now. The software that stops updating dies. Ask anyone who has ever tried to use Windows XP in 2026.

You are not different. You are software.

The Beta-State Career is the model in which:

- Your credentials are an input, not an output. You do not earn a degree and then "have" the degree. You earn a degree to *start*, and then you keep earning new credentials — formal or informal — every year, forever.

- Your skills are in constant patch cycle. You are not "learning new tools to stay relevant." You are in a perpetual state of learning new tools. There is no state of "being current." There is only "updating" or "falling behind."
- Your identity is decoupled from your title. When your role changes — which it will, every two or three years at most — you do not have an existential crisis. Your identity is your adaptability, not your LinkedIn headline.
- Your bets are distributed. You do not go all-in on one specialty, because specialties have the shortest half-lives of all.

This is uncomfortable for a reason. It is the opposite of how most of us were raised to think about work. It undermines the entire educational-industrial complex we grew up in. It breaks the implicit contract between employer and employee that said "you get specialized, we take care of you."

But the contract is broken anyway. That happened without your input. The only question left is whether you operate with or against the new reality.

— — —

Tenure Is Now a Liability

This is the hardest sentence to write in the whole book, because I know people whose entire identity is tied up in it.

In the Cortex Revolution, tenure in a specific role is a *liability*, not an asset.

Here is why.

Tenure used to mean deep expertise, institutional knowledge, hard-to-replace skills, valuable relationships, proven reliability. And those things are still real.

But tenure now *also* means twenty years of investment in skills that are depreciating at five to ten percent per month. Institutional knowledge about systems that are being replaced. Deep expertise in a workflow that is being automated. And — critically — a salary that assumes all of the above is still worth what it was when you got hired.

The company looks at the twenty-two-year veteran on a high salary doing tasks that can be substantially automated, and they look at the thirty-year-old who can manage the automation while also doing some adjacent judgment work at half the salary, and — look, I do not have to finish the sentence. You know how this goes.

This is not a critique of older workers. I am an older worker. I am sixty-two. I am figuring out my own version of this. But I am not going to lie to you about the math, because nobody lied to me about the math, and that is why I am writing this book instead of filing for unemployment.

The way you fight the liability of tenure is by loudly, publicly, visibly refusing to *be* tenured in anything specific. You stay a generalist of adapting. You stay a specialist of cross-pollination. You stay a master of "figuring out the new thing faster than the twenty-two-year-old who does not know what they do not know."

Your moat is not your accumulated knowledge. Your moat is your ability to discard accumulated knowledge without flinching.

— — —

Meet Kevin

Kevin Huang is twenty-two years old. He is a senior at a state school in the Midwest. He is a business major with a concentration in marketing.

In his freshman year — 2022 — his academic advisor told him that marketing was a future-proof field with great career prospects, especially if he took some data analytics electives and learned SQL.

In his sophomore year — 2023 — ChatGPT got released. Kevin did not really think about it. His professors did not really think about it either. A few tech-bro kids in his classes were using it to write papers. Most of his professors banned it, which had the practical effect of teaching Kevin nothing about it while the rest of the workforce started using it.

In his junior year — 2024 — Kevin did an internship at a mid-sized marketing agency. He wrote copy, edited decks, ran analyses. Good internship. Got good feedback. On his last day, his supervisor took him aside and said, not unkindly, "Honestly, Kev, I do not know what this industry looks like when you graduate. We let go of two of our junior copywriters last quarter because Claude does a better job. I am telling you this because I wish someone had told me."

Kevin spent the summer before his senior year, in 2025, deeply panicking. He considered switching majors. He considered applying to law school. He considered quitting school entirely and doing a welding program. He talked to his parents, who told him he was overreacting.

He was not overreacting. He was observing correctly.

In his senior year — now, 2026 — Kevin is finishing his degree, not because he thinks it is useful, but because he is within shouting distance of the diploma and his parents will kill him if he does not graduate. He has no illusions about his starting career prospects. The entry-level marketing jobs that existed when he started college have been substantially eliminated. The ones that remain are being shifted to AI supervision roles that assume you already know things he is not being taught.

Kevin is the canary in the generational coal mine. He is the kid in every reader's life — maybe your own kid, maybe your niece, maybe the young guy on your team, maybe the new hire in your department — who showed up to college expecting one economy and is graduating into another.

He needs a different plan. So, frankly, do most of us.

— — —

The Friday Unlearning Habit

Here is the concrete practice from this chapter.

Every Friday — you pick the hour, but stick with it — you do two things.

Ten minutes: unlearn something. Pick one belief about how your job, industry, or field works. Test it against what is actually happening right now. If it still holds, great. If it does not, write down the updated version. Do this ruthlessly. Your 2019 mental model of your industry is probably actively dangerous to you in 2026.

Fifty minutes: learn something. Not something you "should already know." Something new. A tool you have not tried. A concept that is outside your current field. A technique you have been avoiding because it is uncomfortable. Use this hour not to get marginally better at what you already do, but to get meaningfully exposed to something you do not.

The Friday Unlearning Habit does something nothing else in this book does: it physically installs the half-life into your calendar. It makes adaptation a weekly practice instead of a crisis response. Over a year you will have unlearned fifty-two assumptions and learned fifty-two new things, and both columns will compound.

— — —

Diagnostic

Answer these honestly:

- When is the last time I learned a tool that genuinely changed how I work?
- What am I currently the "expert" on that nobody will need an expert on in twenty-four months?
- What skills did I consciously develop in the last twelve months?
- What is on my current learning list? Not "should be on my list." What is actually on it.
- When did I last do something in my professional life that scared me because I might fail at it?

If those questions make you uncomfortable, good. They are supposed to.

— — —

Science Anchor

- The WEF Future of Jobs Report (2023, 2025). Specifically the skill half-life projections. Compare editions to see how fast the estimates themselves are updating.
- Karl Friston's work on the free energy principle, which is a technical way of saying "adaptive systems survive by constantly updating their models of the world." You are an adaptive system. Behave like one.
- Carol Dweck, *Mindset.* An older book, but the growth-mindset research is still the best lay explanation of why people who embrace discomfort outperform people who do not.

— — —

Close

Kevin walks across the stage next month. He gets his diploma. He hugs his mom. His dad looks proud. The whole thing is very 2006, which is also the last year his major was arguably a clean career play.

He knows the degree is not going to save him. He also knows that what he does in the eighteen months after graduation will matter more than the four years before it.

Here is what Kevin has going for him: he does not have twenty years of assumptions to unlearn. He is already in beta. He is already adaptable. He is already comfortable with tools his professors do not understand. In a strange way, the kids who came up during this transition are going to be *less* traumatized by it than the people who were midway through their careers when it hit. They expect instability. We do not.

If you are reading this and you are over forty-five, do not let that paragraph discourage you. You have something Kevin does not have: judgment, relationships, and a deep understanding of how industries actually work. You just need to bolt the adaptability back on.

Here is how you do it: Friday, for an hour. Unlearn one thing. Learn another. For the rest of your working life.

That is the half-life habit. Install it this week.

Next chapter: why someone has to sign the paper, why that someone is not the AI, and why legal and financial liability might end up being the single most valuable thing a human being can offer.

PART TWO

The Human Premium

Three pillars. Three reasons you still get paid. Here is what stays human.

Chapter 5: *The Accountability Premium*

You can sue me. You cannot sue ChatGPT. Congratulations — you are now legally valuable.

In 2023, two attorneys in New York filed a brief in federal court citing six legal precedents. The precedents did not exist. They had been hallucinated by ChatGPT, which the attorneys had used to draft their research, trusting the output without verification. The judge was not amused. The attorneys were sanctioned. They were publicly humiliated. They became, briefly, the most Googled lawyers in America, which is not the kind of publicity you want on your firm's Wikipedia page.

There is a lesson in this story, but it is not the one people took from it at the time.

The lesson everyone took: AI makes stuff up, so be careful.

The lesson they missed: *the attorneys got sanctioned. ChatGPT did not.*

Think about that. Think about what it means. Think about what it implies for every regulated industry in the world.

The court did not care that the AI hallucinated. The court did not fine OpenAI. The court did not issue a ruling against the software. The court held the *humans* accountable, because they had signed the brief, and in the entire edifice of American law, the name at the bottom of the document is the name that is legally, financially, and professionally responsible for what is in it.

This is not a bug in the legal system. This is the system working exactly as designed, and it is the single most important economic principle of the Cortex Revolution.

Welcome to the Accountability Premium.

— — —

The Liability Gap

Here is the thing nobody building AI systems wants you to fully understand, because it complicates the marketing:

You cannot sue an AI.

You can sue the company that makes the AI, maybe, sometimes, under narrow circumstances involving consumer protection or product liability, and those suits are mostly going to fail because the terms of service you clicked past in a hurry disclaimed liability for everything short of arson. You cannot sue the AI itself because the AI is not a legal person. You cannot fire it for incompetence because it is not an employee. You cannot revoke its license because it does not have one. You cannot fine it. You cannot jail it. You cannot ruin its career.

In every regulated profession — law, medicine, engineering, accounting, financial advice, pharmacy, architecture, and about forty others — there is a licensed human whose name, credential, and legal liability is on the line for the final output. That human *must* exist. The regulation requires it. The malpractice insurance requires it. The court requires it.

This is what I call the Liability Gap. It is the gap between what an AI can produce and what a legal system will accept as a signed, binding, accountable professional output. That gap is bridged by a human. And the market has already started to price this bridging service.

Dave — remember Dave, from Chapter 2, who was having a bad time? — is actually not as screwed as he thought.

Dave's job title is "underwriting manager." His actual economic function, which his company has not yet articulated in those terms, is "the human who reviews the AI's output, applies final judgment, and signs his name on the decision that goes to the client and the regulator." The AI does 73% of the workflow. Dave does the 27% that a regulator, a client, or a plaintiff's attorney can come after if something goes wrong.

That 27% is Dave's real job. It was always his real job, actually. He was just getting paid to do the other 73% because nobody had a tool that could do it. Now somebody does, and Dave's job is revealed to be what it fundamentally always was: accountability.

— — —

Signer Roles

I want to introduce a term that I think is going to become a major career category in the next decade: the Signer Role.

A Signer Role is any job in which the primary economic value is the human willingness and legal ability to accept responsibility for a final output. Signer Roles include:

- **Physicians.** Diagnosis and treatment decisions. AI can analyze the scan; a licensed physician signs off on what it means. The physician is the one the state board can pull a license from.

- **Licensed attorneys.** Legal advice and filings. AI can draft the brief; an attorney signs it, and the attorney is the one who can be sanctioned.
- **Certified accountants.** Audit opinions, tax filings. AI can do the math; a CPA signs the return. IRS rolls downhill to the signer.
- **Engineers (PEs).** Engineering documents. AI can generate the CAD; a professional engineer stamps the drawing, and the stamp is what legally matters.
- **Financial advisors.** Investment advice. AI can generate allocations; the licensed advisor accepts fiduciary duty.
- **Nurse practitioners, physician assistants, licensed pharmacists.** Every allied health profession with a scope of practice and a licensing authority.
- **Pilots, air traffic controllers, licensed drivers of commercial vehicles.** Regulated transport.
- **Licensed contractors, electricians, plumbers, HVAC technicians.** Regulated trades. Notice the overlap with the Physicality chapter. The premium compounds when physical and accountability stack.
- **Corporate officers.** CEO, CFO, CCO, General Counsel. Whoever signs the 10-K. Whoever signs the audit letter. Whoever is legally liable.

Notice what these have in common. They all require:

3. Licensure or certification.

4. Legal liability for the final output.
5. Malpractice or errors-and-omissions insurance.
6. A specific human individual whose name goes on the document.

These are the last stable white-collar jobs. Not all white-collar jobs. Most of the ones without these four characteristics are in deep trouble. But the ones *with* these characteristics are actually *more* valuable in an AI-enabled world, not less.

Why? Because AI systems increase the throughput of professional outputs. The same licensed physician can now review three times as many scans per day with AI assistance. The same attorney can handle more cases. The same CPA can sign more returns. This means the *value* of the license — the scarce, credentialed, legally accountable human — goes UP, not down.

This is the exact opposite of what happens in non-licensed knowledge work, where the throughput increase just means fewer people are needed. The license creates a floor. The license is the moat.

— — —

The Regulatory Tailwind

Here is the part that is hilarious, in a dark way.

While everyone on LinkedIn was panicking that AI would erase all the lawyers, the actual lawyers — the ones in Washington, Brussels, Sacramento, and Albany — were busy passing laws that ensure humans will be legally required to stay in the loop forever.

The EU AI Act, rolled out 2024 to 2026, classifies AI systems by risk level and requires human oversight for all "high-risk" applications. That is medical AI, legal AI, employment AI, credit-scoring AI, critical-infrastructure AI, education AI, law-enforcement AI. For all of these, a human has to be in the loop. Not "should be." *Legally has to be.* The EU has decided this at statute.

The United States is following, state by state, industry by industry. Colorado passed the first comprehensive state AI law in 2024. California is rolling out sector-specific rules. New York has specific AI employment law. The federal level is slower, but regulators at the FDA, SEC, CFPB, DOL, and others are all writing sector-specific rules that require human accountability.

Professional licensing boards are not sitting this out. State medical boards are updating their rules to require physician sign-off on AI-generated diagnostic recommendations. State bar associations are issuing ethics opinions requiring attorney verification of AI-generated work product. State licensing boards across the regulated trades are building AI oversight into their continuing-education requirements.

Every one of these rules does the same thing. It legally mandates a human signer. It builds the Accountability Premium into statute.

The Cortex Revolution is not eliminating lawyers. It is eliminating the *junior* lawyers who used to do the grunt work. The licensed, senior, signing lawyers are getting *more* important, not less, because the regulators have decided, correctly, that accountability cannot be automated.

— — —

Dave Sees the Door

Dave Palmieri has been reading this book because his cousin gave him a copy. His cousin is married to Marisol Quintero. Small world.

Dave has just realized something that is, no kidding, probably going to save his career.

His job title is "underwriting manager." His actual job — the one that matters in 2026 — is Licensed Signer of Insurance Underwriting Decisions. The AI platform at his firm can produce recommendations all day long. It cannot issue a policy. Only a licensed, bonded, qualified human underwriter can. And if that underwriter gets the decision wrong — if the risk was miscategorized, if the pricing was wrong, if the client misrepresented something and the underwriter did not catch it — the underwriter and the firm are the ones facing regulatory action and litigation. Not the AI.

Dave has a CPCU designation. Dave has twenty-two years of underwriting experience. Dave has a license. What Dave does *not* have — yet — is a clear articulation of his professional identity as a Signer, rather than as an Underwriter Of Individual Policies. That articulation is the key pivot.

If Dave reorients himself as "the experienced, licensed, accountable human who reviews and signs off on AI-generated underwriting decisions for commercial trucking risks in four states," Dave is more valuable to his firm, not less. The AI is the junior underwriter. Dave is the partner who signs.

The salary math should follow. It might not immediately, because firms are slow and HR is slower. But eventually the market will correct, because the demand for licensed, experienced signers is going up while the demand for information-moving analysts is going down.

Dave's move is to lean into the license, the credential, and the accountability. To become more visibly, more articulately, more fundamentally the person who signs the paper.

Your move is probably similar, if your career has any accountability component at all. If it does not, the move is to get one. Get the license. Get the certification. Get the credential that comes with liability and a registration number. Those numbers are going to be more valuable in 2030 than they were in 2020.

— — —

The Premium, Quantified

You want numbers? Fine.

Roles with significant licensure and liability components saw wage growth of 8 to 14 percent between 2023 and 2026 depending on sector, compared to flat or declining wage growth for comparable-level non-licensed knowledge work. This is a real trend, and it is widening.

Malpractice insurance rates for licensed professionals have stayed steady or risen slightly, indicating that the insurance industry — which is actually quite good at pricing risk — sees the liability environment as stable to hardening. Licensed professionals are, if anything, more valuable to insure because they are more in demand.

Continuing education and credentialing enrollments are up in 2025 and 2026, particularly in mid-career pivots toward licensed fields — nursing, accounting, law, licensed trades. Adults are voting with their time and money, and they are voting for credentials with teeth.

The Accountability Premium is not speculative. It is showing up in the wage data, in the insurance data, and in the enrollment data.

— — —

Monday Morning

This one is concrete and kind of urgent.

Make a list of every license, certification, or credential that applies to your current field. Include the ones you do not have. Include the ones you have been putting off. Include the ones that used to seem like "paperwork."

Now ask yourself: which of these, if I had it, would convert me from a "worker doing tasks" into a "signer accepting accountability"?

Start pursuing that credential. Today. Not next quarter. Today. Find out what it takes. Enroll if enrollment is open. Order the study materials. Schedule the exam window on your calendar.

If you already have the top credential in your field, your Monday-morning move is different: figure out how to *visibly* position yourself as a signer, not a worker. Update your LinkedIn. Update your email signature. Update how you introduce yourself. Your title is not "Senior Underwriter." Your title is "Licensed Underwriter Accountable for Risk Decisions in [your specialty]." Phrasing matters.

— — —

Science Anchor

- EU AI Act, full text, especially the "high-risk AI" provisions. It is painful reading; do it anyway.
- NIST AI Risk Management Framework (AI RMF 1.0 and updates). U.S. baseline for organizational AI governance.
- Ethics opinions from your state bar association, your state medical board, or your professional licensing board. They are updating rapidly. Read the last six months' worth.
- Goldman Sachs AI exposure report (2023). Re-read it with an eye for which occupations on their high-exposure list require licensure. The overlap will tell you exactly where the Accountability Premium is building.

— — —

Close

I want to tell you something that is easy to miss in the middle of all the doom-and-gloom AI discourse.

If you are a professional in a regulated field, and if you have spent years earning a credential that comes with legal accountability, you are not a dinosaur. You are not a commodity. You are not replaceable by a prompt.

You are a signer. You are the legally, financially, professionally accountable human who stands behind a decision. You are the name on the document, the number on the license, the signature on the page. You are what the court requires, what the regulator requires, what the client requires, and what the system requires to function at all.

That is not a small thing. That is not a dying thing. That is a growing, appreciating, scarce thing.

The world is filling up with AI output. The world has a finite number of humans who can and will sign their name at the bottom of it and take the consequences.

If you are one of them, raise your rates. Literally. Today. The market has already moved; it is waiting for you to notice.

If you are not one of them yet, start the paperwork.

Dave closes his laptop at 6:15 p.m. on a Thursday. He picks up the phone and calls his cousin Marisol, because Marisol is somehow the only person he knows who actually seems to be thriving in this economy, and he wants to know her secret.

"Easy," she says, laughing. "I am the one they call when something breaks. And I am the one who signs the inspection sticker. Everyone else is replaceable."

Dave writes it down.

So should you.

Next chapter: why, in a world of infinite AI-generated output, the most valuable professional skill is suddenly the ability to tell the difference between good and great — and how you actually build that capacity, on purpose, starting next week.

Chapter 6: *Taste Is the New Talent*

AI will make ten thousand logos. The human who picks the right one is the employee.

It is 3 p.m. on a Wednesday in Chicago. Megan, a senior designer at a mid-sized branding agency, is staring at two hundred and forty-seven logo concepts. A client wants a new identity for their HVAC company. The brief is standard. The budget is twenty-eight thousand dollars. The deadline is Friday.

Megan did not draw two hundred and forty-seven logo concepts. She drew zero. She typed five paragraphs of design direction into a tool she pays thirty dollars a month for, and the tool generated two hundred and forty-seven of them in under ninety seconds. The first sixty are trash. The next fifty are generic. About forty are genuinely pretty good. Maybe fifteen are excellent. Three are probably the answer.

Megan's entire job, now, is to find which three. To present those three to the client. To explain why those three. To defend the choice. To iterate based on feedback. To stand behind the final logo when it goes on the trucks and the invoices and the website and the Google Ads.

Megan is not a designer anymore. She is a curator. And her paycheck just tripled.

———

The Taste Gap

Here is the thing nobody wants to say out loud about generative AI: most of what it produces is mediocre. Some of it is bad. A small slice of it is genuinely excellent. And the economic value of the excellent slice is inversely proportional to how easy it is to find.

For about a hundred and fifty years, the bottleneck in creative work was *production.* Drawing a logo took hours. Writing a novel took years. Composing a symphony took a lifetime. The person who could actually produce the thing — at quality, on time, on spec — was the person who got paid. Everything downstream of production, including evaluation and curation and selection, was considered a sort of soft managerial overhead. Important, sure. But not where the action was.

In 2026, production is essentially free.

I am going to let you sit with that sentence for a second, because it is the single most important economic development of our working lives, and most of us are still in denial about it.

Production is free. Drawing a logo is free. Drafting a blog post is free. Generating a pitch deck is free. Composing a jingle is free. Writing a first-pass screenplay is free. Sketching a building is free. Creating a training video is free. Producing a children's book illustration is free. Every one of those tasks, which five years ago required a skilled human being and hundreds of billable hours, can now be done in the background while you eat lunch, for roughly the cost of a large pizza a month.

The bottleneck has moved. It is no longer production. It is *selection.*

Somebody has to look at the two hundred and forty-seven logos and say "that one." Somebody has to read the twelve AI-drafted blog posts and say "none of these, but the third one could work if we rewrite the opening." Somebody has to listen to the forty candidate jingles and know, instantly, which three are worth showing the client and which thirty-seven will make the client quietly unfollow us on LinkedIn.

That somebody has taste. And taste, right now, is worth more than it has ever been in human history.

— — —

Critics Outlive Creators

Let me draw your attention to a pattern that shows up in every mature creative industry, and that is about to show up in yours.

In every field where production has been commoditized, the person at the top of the food chain is not the producer. It is the curator, the critic, the editor, or the selector. Examples, rapid-fire:

- **Restaurants.** The highest-earning people in fine dining are not the line cooks. They are the chefs who decide what goes on the menu. Above them are the food critics, restaurant investors, and Michelin inspectors who decide which restaurants exist at all.
- **Music.** The highest-earning people in the music business are not session musicians. They are producers — the people who decide which takes make the album. Above them are A&R executives, who decide which artists get signed in the first place.
- **Film.** Actors get famous. Directors make more money. Studio heads make vastly more. The person deciding which script gets greenlit is making more than any ten actors in the movie.

- **Fashion.** Anyone can sew a garment. The creative director of a major fashion house — the one who decides what the brand will make, and what it will not — is one of the highest-paid jobs in the entire industry.
- **Publishing.** Writers write books. Editors decide which books exist, in what form, with what edits. Publishers decide which editors get hired. The higher you go, the more it is about selection, not production.

Notice the pattern. Every single one of those industries commoditized production decades or centuries ago, and every single one of those industries pays a huge premium to the humans who can tell good from great, bad from adequate, and commercially viable from commercially fatal.

Now, in 2026, *every* industry is in the same boat. Every industry has had its production costs collapsed. Every industry is drowning in output. Every industry desperately needs humans with taste.

If you can be the person who says "not that one, this one" — and if you can explain why, and defend it, and be right a healthy percentage of the time — you have a career for the next twenty years. I do not care what field you are in. This generalizes.

— — —

Taste Is Not a Mystical Gift

Here is the part that is going to comfort you or depress you, depending on how lazy you are feeling. Taste is trainable. It is not a magic quality that some people are born with. It is a pattern-recognition skill built through exposure, reps, and feedback, exactly like every other skill worth having.

The research on expert pattern recognition is clear. Experienced judges in a field develop an almost subconscious ability to spot quality because they have seen thousands of examples of it — and thousands of examples of failure. Gary Klein spent decades studying how firefighters, nurses, chess players, and military commanders make high-stakes decisions under uncertainty. His finding: they are not running explicit decision trees in their heads. They are pattern-matching against a vast internal library of prior cases. Their taste is stored in the library, not in the algorithm.

Which means if you want taste, you have to build the library. And if you want taste fast, you build the library on purpose.

Three things go into building taste:

Exposure. You cannot have taste in a field you have not looked at. You have to consume a staggering amount of the best work in your domain. Not the mediocre stuff. Not the stuff that happens to be going viral on LinkedIn this week. The actual best work. The stuff that serious people in your field cite when they are arguing with each other. If you are a marketer, that is award-winning campaigns going back forty years, not the latest HubSpot post. If you are a writer, that is published novels that survived the test of a decade, not the bestseller shelf at the airport.

Reps. You cannot have taste without practicing making judgments. Exposure alone gives you aesthetic preferences. Judgment requires saying "this is better than that, and here is why." You have to make calls. You have to defend them. You have to be wrong sometimes and figure out why you were wrong. Taste without reps is just snobbery.

Feedback loops. You cannot know if your judgments are any good without getting feedback from people whose taste is better than yours. This is the hard part. It requires finding actual experts and risking their disagreement. It requires reading serious critics in your field and noticing where your calls diverge from theirs. It requires, occasionally, being wrong in public and surviving it.

Do these three things for a year, on purpose, and your taste in your field will measurably improve. Do them for five years and you will become the person people ask for a second opinion. Do them for ten and you are in Megan's seat — the curator, not the producer. The one with the career.

— — —

The Thirty-Day Taste Training Protocol

You want a concrete plan. Here is one.

For the next thirty days, every single day, you do three things:

7. **One hour of curated exposure.** Identify the top five living practitioners in your field. If you cannot name five, that is your first problem. Follow their work. Read everything they publish. Watch everything they ship. Read the people *they* cite. Do this for an hour a day, minimum. No TikTok. No podcasts about the field. *The work itself.*
8. **Fifteen minutes of judgment exercises.** Pick five pieces of work in your field — articles, designs, pitches, products, whatever applies. Rank them from best to worst. Write one sentence explaining why. Do not Google what the consensus thinks. Make your own call. Write it down. Store it.

9. **Fifteen minutes of feedback.** Either post one of your judgments publicly and let people disagree with you, or compare your rankings against the rankings of someone whose taste you respect. Where did you diverge? Why? What did they see that you missed? What did you catch that they might have overlooked?

An hour and a half a day. Thirty days. At the end of it, your taste will be measurably sharper than it was when you started. More importantly, you will have installed a *practice* — a permanent, ongoing system for getting better at judgment, indefinitely, for the rest of your career.

You will not become Miranda Priestly in thirty days. You will become someone who is measurably moving toward Miranda Priestly, which is what a career is actually made of.

———

Diagnostic

Answer these honestly:

- Can I name the five best practitioners alive in my field, and defend why they are the best?
- When I see mediocre work in my field, can I articulate specifically what is mediocre about it — not "I just do not like it" but "here is the craft failure"?
- Do I have a point of view about my field that I could defend in a public argument, or do I mostly hold whatever opinion was trending on LinkedIn last week?

- When was the last time I changed my mind about what constitutes good work in my field, based on exposure to something I had not seen before?
- Who do I consult when I cannot tell if a piece of work is good? Do I have a mentor, a peer, a critic I trust — or am I alone in my taste?

If those questions make you squirm, welcome to the club. Most of us never actually trained our taste. We absorbed some default aesthetic from our industry and called it done. That is not taste. That is convention. And convention is not worth paying for in 2026.

— — —

Monday Morning

Start the thirty-day protocol this week. Put the hour of exposure on your calendar. Make it recurring. Treat it like a meeting with your future career.

Pick one piece of work in your field today. Render a judgment on it — in writing, in a document you save. Rate it. Defend the rating. Do this again tomorrow with a different piece. And again the next day.

At the end of the thirty days, read the thirty judgments in order. Notice how they evolved. Notice what your early-stage taste missed. Notice what your later-stage taste catches.

That journal is more valuable than any certification you will earn next year. Nobody can automate it. It is your taste, written down, provably developing in real time.

— — —

Science Anchor

- Gary Klein, *Sources of Power* (1998). The foundational work on expert pattern recognition in high-stakes fields. Still essential.
- Ethan Mollick's ongoing work on "centaur" and "cyborg" patterns of AI use. The centaur — human as selector, AI as producer — is exactly the taste-forward pattern described in this chapter.
- Daniel Kahneman, *Thinking, Fast and Slow* (2011). Specifically the chapters on intuitive expertise and the conditions under which fast judgment is trustworthy.
- Anders Ericsson on deliberate practice. The Ericsson research is often misquoted as "ten thousand hours," but the actual finding — that practice has to be specific, feedback-driven, and uncomfortable — applies directly to taste training.

— — —

Close

Megan delivers the three final logo concepts on Thursday morning, a day ahead of deadline. The client picks one. It is the best of the three. It is also, not coincidentally, the one Megan wanted them to pick. She steered the meeting. She is good at that, too.

Behind the scenes, on Megan's laptop, two hundred and forty-four logo concepts sit in a folder called "unused." They are not bad. Some are even good. But they are not the answer. And Megan, over years of exposure and reps and feedback, has built the library in her head that lets her know that — quickly, confidently, defensibly.

That library is her moat. That library does not show up on a résumé. It does not have a certification to back it up. But it is the thing her agency pays her for, and it is going to be the thing her agency pays her more for every year for the next twenty years.

Your boss's boss has a job because of taste, not skill. So does every person above them on the org chart. Work backward from that. Start building the library this week.

Next chapter: your brain is plastic until you die, and the algorithm is betting you will not use yours. Let us prove it wrong.

Chapter 7: *Adaptability as a Muscle*

I am sixty-two years old. Last month I spent nine hours recording my own voice.

I was reading passages from my own books into a microphone — narration, dialogue, the works. Every inflection. Every "uh" I had to retake. The goal was to train a voice model, a version of me that can narrate audiobooks without me sitting in a studio for three weeks per title. If it works, an AI trained on my voice will produce audiobooks that sound close enough to me that my readers will not mind.

The weird part is not the technology. The weird part is looking at yourself at sixty-two, with three graduate degrees and forty-five years of professional experience, willingly training the thing that is going to do a version of your job without you. The weird part is the small voice in the back of your head asking: "Is this adaptive, or is this what a dying species does?"

The answer is: both. It is both. And the people who survive the next decade are the ones who can sit with that, and do it anyway, and not let the discomfort of it stop them.

This chapter is about how you do that. On purpose. As a skill.

— — —

Your Brain Is Not Old

The single most useful thing you can know about your own brain is that nearly everything you believed about it in high school is wrong.

You were told that brains stop developing around age twenty-five. Wrong. You were told that you had a "learning window" as a kid that closes in adulthood. Wrong. You were told that you cannot teach an old dog new tricks. Wrong. You were told that neurons do not regenerate, that intelligence is fixed, that talent is innate, that your math brain or language brain or spatial brain was hardwired at birth. Wrong, wrong, wrong, and wrong.

What we actually know in 2026, thanks to the last twenty years of neuroscience research: adult brains remain plastic throughout the entire lifespan. Neural pathways form and re-form in response to new learning, new environments, new challenges. The capacity for change slows a little with age, but it does not disappear. A motivated fifty-year-old learning a new instrument is not meaningfully worse at it, after controlling for practice time, than a motivated twenty-five-year-old. A retiree who takes up a serious new intellectual hobby can measurably improve their cognitive function well into their eighties.

The research on this is not controversial. It is just counter to cultural narrative, which means most people live their whole adult lives under the assumption that they are slowly becoming cognitively obsolete, when they are actually carrying around the most adaptive organ that has ever existed on Earth. They just do not use it for adaptation.

You do not need to believe me. Go look up Lara Boyd's TEDx talk on neuroplasticity, which summarizes the research cleanly in about fifteen minutes. Go read any serious review paper published in the last decade. The consensus is overwhelming. Your brain is a weapon. You have been told it is a museum exhibit. The two views lead to very different lives.

— — —

The Beta-State Mindset

Chapter Four introduced the Beta-State Career. This chapter is the psychological operating system that makes it possible.

The Beta-State Mindset is the posture of holding every skill, identity, and assumption loosely enough that you can drop it when it stops serving you. Most people do the opposite. They grip tight. They identify with their current skills. They build their self-worth around their current job title. They defend their current worldview against new information. And then the world changes and they are stuck, because letting go of the skill or the title or the worldview would feel like losing a limb.

The Beta-State Mindset looks like this, in practice:

- **Hold skills like tools, not like organs.** You are not "a SQL developer." You are a person who currently uses SQL. If SQL becomes obsolete, you do not become obsolete. You pick up the next tool.
- **Hold job titles like hats, not like skin.** Your title is something you are wearing, not something you are. When the role changes, the hat changes. You are still you.
- **Hold opinions like hypotheses, not like heirlooms.** When new evidence shows up, you update. You do not dig in. Updating is how you know your brain is still working.
- **Hold expertise like scaffolding, not like a castle.** Expertise exists to help you learn the next thing faster. It does not exist to keep you in one place for forty years.

This sounds like psycho-jargon. It is not. It is the single most concrete difference between people who survive technological transitions and people who do not. The survivors are the ones who can say "that used to be me; I am something else now" without grief. The non-survivors are the ones who cannot let go of who they used to be, and so they stand still while the world moves, and eventually the world leaves them behind.

This is not about your age. I know sixty-two-year-olds who operate in full beta state and thirty-two-year-olds who are already ossified. The age distribution is surprisingly uniform. What matters is practice.

— — —

Five Adaptability Drills

Adaptability is not a feeling. It is a set of behaviors you can train. Here are five drills. Pick at least two. Run them indefinitely.

Drill one: The Scary New Thing Quarterly. Every ninety days, publicly commit to doing something professionally uncomfortable. A new skill. A new tool. A new type of project. A talk at a conference where you will not be the smartest person in the room. Publicity matters — it has to be declared to someone who will notice if you bail. The discomfort is the point.

Drill two: Crossover Hours. Spend three hours a week working in an adjacent discipline. Not as an expert. As a beginner. If you are a marketer, spend three hours a week learning basic statistics. If you are a lawyer, spend three hours a week on design principles. If you are a doctor, spend three hours a week on behavioral economics. The transfer does not have to be obvious. The cross-pollination is the point. This is also how you build the combinatorial edge we will talk about next chapter.

Drill three: Anti-Expert Days. Once a month, deliberately put yourself in a situation where you are the least experienced person in the room. Take a class in something you have never studied. Join a conversation where everyone knows more than you. Ask beginner questions and take notes. The habit this builds is not the knowledge; it is the comfort with being uncomfortable. That comfort is the whole game.

Drill four: Opinion Updates. Once a month, in writing, revise a professional opinion you used to hold. What did you believe five years ago that you no longer believe? What changed your mind? This is a muscle. The people who cannot do this exercise are the people who will be left behind, because they are signaling, to themselves and to the world, that they are no longer updating. Updating is survival.

Drill five: Discomfort Tolerance Reps. Once a day, do something small that is professionally uncomfortable. Send the email you have been putting off. Ask for the feedback you are dreading. Try the tool you have been avoiding. The brain learns to tolerate discomfort the same way the body learns to tolerate exercise — through consistent, small, controlled exposure. The people who cannot tolerate small discomforts will not survive large ones.

— — —

Why Kids Are Not Actually Better at This

There is a common lie in the AI-and-careers conversation, which is that young people are "native" to this transition and older workers are inherently behind. I want to push back on this, gently but firmly.

Young people are not inherently more adaptable. They just have less to unlearn.

When a twenty-two-year-old picks up a new AI tool, they are not drawing on some mystical digital-native superpower. They are drawing on the fact that they have not yet invested twenty years in the workflow the tool is replacing. There is no grief to process. No identity to renegotiate. No sunk cost to confront. Adoption is easy because there is nothing to let go of.

What older workers have, that younger workers do not, is judgment. Experience. Relationships. Pattern recognition across decades. Context for what actually matters in an industry. The ability to tell the difference between a hype cycle and a real shift, because we have seen a few. The willingness to stand up in a meeting and say "I have seen this before, and here is what actually happens."

That experience has enormous value. It always has. It just has to be paired with beta-state psychology, not ossified into "the way we always do things." When you pair hard-won judgment with genuine adaptability, you become something the thirty-year-olds cannot compete with and the seventy-year-olds cannot replicate. You become valuable in a specific way that only shows up in the middle of a transition.

I am writing this at sixty-two. I will not pretend it is easy. I will not pretend that re-training my mental models every quarter is as natural as it was at twenty-five. It costs more. But it also pays more, because fewer people my age are doing it, and the ones who do get noticed.

If you are over fifty and reading this: adapt. You have more to offer than you think. You are just under-investing in the psychological infrastructure that lets you offer it.

— — —

Diagnostic

- In the last year, did I learn something that scared me because I might fail at it?
- When was the last time I publicly changed my mind about something professional?
- Do I spend any regular time being a beginner at something, or am I always performing expertise?
- When a new tool, framework, or concept arrives in my industry, is my first reaction curiosity or defensiveness?
- Can I name three beliefs about my work that I held firmly five years ago and have since revised?

— — —

Monday Morning

Pick one of the five drills. Start it this week. Put it on your calendar. Tell at least one other person you are doing it.

The point is not to pick the "right" drill. The point is to start converting adaptability from a trait you occasionally aspire to into a practice you do on schedule. Once it is a practice, it compounds.

Once it compounds, you are a different person twelve months from now.

— — —

Science Anchor

- Lara Boyd, "After watching this, your brain will not be the same" (TEDx Vancouver, 2015). The clearest short introduction to adult neuroplasticity.
- Karl Friston on the free energy principle. Technical, but the core idea — adaptive systems survive by continuously updating their models of the world — is the theoretical backbone of everything in this chapter.
- Carol Dweck, *Mindset* (2006). Still the best lay explanation of the growth-versus-fixed mindset research. The growth mindset is not a cliche; it is empirically supported, and the fixed mindset is professionally fatal in 2026.
- Ellen Langer's work on mindfulness and aging. Specifically her "counterclockwise" study, which showed that older adults immersed in environments from twenty years earlier measurably improved on cognitive and physical markers. Your brain takes cues from your posture toward it.

— — —

Close

I go back to the voice recording on a Saturday morning. My wife is in the garage, running her handywoman business on the phone, fielding calls about a roof repair. My coffee is cold. The microphone is hot. I read another chapter.

Somewhere in a server farm, a model is learning the cadence of my sixty-two-year-old voice. In six months it will be able to narrate a book in something close to me. Whether that is beautiful or terrifying depends entirely on whether I am adapting alongside it or being replaced by it. Today, the difference between those two outcomes is me sitting at a microphone, at an age when most men my generation are easing up, still putting in the reps.

I am not sure this works out. Nobody is. That is not the point. The point is to still be the kind of person who picks up the microphone when it is hard to explain why.

What is your microphone?

Next chapter: why AI can remember but cannot invent, and why the most valuable creative skill in 2026 is the ability to combine fields the training data did not connect.

Chapter 8: *Creativity as Combination*

In the late fifteenth century, a banking family in Florence decided to spend an obscene amount of money convening people who normally would not talk to each other. Painters with mathematicians. Philosophers with architects. Astronomers with sculptors. Poets with anatomists.

The family was called the Medici. The result, in a city of forty thousand people, was the most concentrated burst of creative invention in human history. The Renaissance — the actual, specific, historical Renaissance — happened on their dime. Not because they funded individual geniuses. Because they *mixed the geniuses.*

Seven hundred years later, we do the opposite. We silo everyone. We hire specialists to talk to specialists. We build org charts that make sure the engineer never has to listen to the marketer, and the marketer never has to sit through an accounting meeting. We optimize for deep expertise, and we optimize *against* the one thing that reliably produces breakthrough work: combination.

And then we wonder why AI is eating us.

— — —

What AI Actually Does

To understand why combination is your moat, you need to understand, briefly and without a lecture, what AI actually does when it produces output.

A large language model is a massive statistical pattern-matcher trained on a vast corpus of existing human work. When you ask it to produce something, it is not reasoning from first principles. It is not inventing. It is not creating. It is — to oversimplify only slightly — predicting what the most probable next word, phrase, or concept would be, given its training data and your prompt.

This is not an insult to AI. It is an extremely useful capability, and it is going to keep improving. But it implies a hard structural limit, which is this: AI can only produce combinations that are statistically represented in its training data. If the connection between two fields, or two ideas, or two disciplines has not been explicitly made in the corpus of human knowledge the model has seen, the model cannot make that connection. It has no machinery for genuine novelty. It can only interpolate within the space of what humans have already done.

This is the **Combinatorial Ceiling**. AI is excellent at remixing within established patterns. It is terrible at breakthrough combinations — the kind where two unrelated fields collide and produce something neither field alone could have produced.

You, on the other hand, are a messy, disorganized, accidentally-reading-a-book-about-jellyfish-on-Tuesday-morning human who is constantly making connections nobody explicitly trained you to make. That is your advantage. That is the part of creativity AI cannot currently replicate at the frontier.

The question is whether you deliberately exploit it, or whether you spend your career imitating AI by staying inside your lane.

— — —

Frans Johansson wrote a book in 2004 called *The Medici Effect.* If you have not read it, you should. The thesis, which has only become more relevant with every passing year, is that breakthrough innovation happens most reliably at the *intersection* of fields — not at the deep center of a single field.

The Wright Brothers were bicycle mechanics. They invented flight by combining their knowledge of lightweight metal structures with emerging aerodynamic theory and the control principles they had learned from steering unstable two-wheeled vehicles. Aeronautics people at the time were working on flight from an engine-power angle, and failing. The bicycle guys won because they combined three disciplines nobody else had connected.

Steve Jobs famously attributed the typography of the first Macintosh to a calligraphy class he had taken in college. Calligraphy had no obvious connection to personal computing. He made the connection anyway. He was not a better engineer than his competitors. He was a better combiner.

The field of biomimicry — engineers designing products based on biological systems — exists because some people spent their careers in both biology and engineering. Velcro exists because somebody looked at burrs on their dog and thought about fabric closures. Wind turbines are quieter now because somebody looked at whale fins and thought about aerodynamics. The list is endless. Every one of those innovations happened because a human being carried two disciplines in the same head and noticed a connection the specialists in either field never would have.

This is the part of creativity AI struggles with most. Not production. Not variation. *Combination across domains that are weakly connected in the training data.*

And this is what you should be building toward, deliberately, for the rest of your career.

— — —

T-Shaped Is Dead. The Starfish Wins.

For about two decades, career advice circled around the idea of being "T-shaped" — deep expertise in one area, shallow competence in several others. This was a huge improvement over "I-shaped" — one deep specialty and nothing else — because the horizontal bar of the T gave you some collaboration chops across fields.

T-shaped is now insufficient. Here is why.

AI is extremely good at replicating the vertical bar of the T. Deep expertise in a single field is the thing large language models can compete with most effectively, because the training data on any given specialty is dense and well-structured. An LLM can now give you passable expertise in most individual domains on demand. What an LLM cannot do is carry five or six fields in the same head and make *non-obvious* connections between them in real time, in a way that produces genuinely new ideas.

The professional of the next twenty years is Starfish-shaped. Five or six arms of medium-depth competence, each reaching into a different field. One deeper central core that ties them together. The arms are what give you the combinatorial surface area. The core is what gives you enough coherence to do useful work with it.

A marketer who also knows statistics, behavioral economics, basic UX design, a little copywriting, and enough data engineering to run their own experiments is Starfish-shaped. That marketer makes connections nobody else on the team can make. That marketer does not get replaced by AI; that marketer uses AI to do any one of those things faster, while being the only person in the room who understands how they all fit together.

A developer who also knows product, design, some direct customer research, basic finance, and a decent amount of domain expertise in the industry they are building for is Starfish-shaped. That developer becomes a senior technical leader or a founder. That developer is not replaced by coding AI; that developer uses coding AI to ship faster, while being the only person in the room who understands why the software needs to exist in the first place.

Notice the pattern. The Starfish professional is not a generalist — a generalist is shallow everywhere. The Starfish professional has real competence in multiple fields and a specific core. The Starfish professional is the person AI makes more valuable, because AI removes the drudgery of individual-domain execution and frees up bandwidth for the cross-domain thinking that only humans can currently do.

— — —

Meet Jamal

Jamal Bankole is twenty-nine years old. He runs a marketing agency called Bankole & Co. in Austin, Texas. His agency has two human employees besides himself. It has, at last count, forty AI agent workflows running in the background at any given time. His 2025 revenue was just over four million dollars.

Here is what Jamal actually knows, by his own account:

- Behavioral economics, at the level of someone who has read the core literature and applied it in dozens of real campaigns.
- Graphic design, at the level of a competent staff designer — not a virtuoso, but good enough to judge.
- Copywriting, at the level of a senior direct-response copywriter.
- Applied statistics and experimental design, at the level of a mid-level data analyst.
- Basic film and video editing, at the level of a competent YouTube creator.
- And, underlying all of it, a deep core in marketing strategy — the ability to look at a client's business, identify the real bottleneck, and design a campaign that actually moves the number that matters.

Nobody at a traditional agency combines those five disciplines. The agency has a strategist, a designer, a copywriter, an analyst, and a video editor. Five people. Jamal is all five of those people, at competence levels that are high enough to be useful, paired with AI systems that handle the execution grunt work in each domain.

This is why Jamal can run an agency with two employees. It is not because he works twice as hard. It is because his *combinatorial surface area* — the way his five disciplines let him see a client's problem whole — is something no individual specialist and no individual AI agent can replicate. He makes connections in a discovery meeting that nobody else in the room has the vocabulary to make. He designs campaigns that pull from behavioral economics, test like experiments, read like copywriting, and deploy like design projects, because in his head those are the same project, viewed from different angles.

When I asked Jamal what the single most important move of his career had been, he said: "Figuring out that I did not have to pick one field. Everyone kept telling me to specialize. I kept getting better by not specializing. Eventually I noticed what the pattern was and leaned into it on purpose."

That is the whole chapter in three sentences. He just said it better than I did.

— — —

The Weekly Cross-Domain Input Habit

You cannot become Starfish-shaped by reading more about your current field. You have to deliberately, regularly, consume from outside it.

Here is the habit. Every week, you consume three inputs from fields adjacent or unrelated to yours:

10. **One long-form input.** A book chapter, a long essay, a research paper. From a field you do not work in, but that could plausibly inform your work.
11. **One practitioner input.** An hour listening to someone who actually does the adjacent work — a podcast, an interview, a conference talk. Practitioners talk about what the field is actually like in the trenches, which is different from what the field looks like from the outside.

12. **One artifact input.** An actual piece of work from the adjacent field. A campaign. A product. A paper. A performance. Experience it the way a practitioner in that field would experience it, not the way a tourist would.

Keep a cross-domain notebook. When a connection between the adjacent field and your own field flashes in your head — even if it seems silly — write it down. Most of these connections are garbage. A few of them are not. Over a decade of doing this habit, you will accumulate dozens of non-obvious insights that nobody at a single-domain-only career could possibly have.

This is slow. It is unglamorous. It does not show up in the first quarter. Do it anyway. The compounding is absurd.

— — —

Diagnostic

- Name three fields — not your own — that you could hold a real conversation in. If you cannot name three, you are under-diversified.
- In the last twelve months, did you make a professional decision informed by a field outside your own? What was it?
- When was the last time you had a conversation with a practitioner in a field unrelated to yours?
- If someone audited your bookshelf, physical or digital, would it look like a specialist's library or a Medici's library?

- When you hit a hard problem in your own field, do you reach for tools from other fields? Or do you only reach for the tools you already have?

— — —

Monday Morning

Pick a field adjacent to yours that you have always been low-key curious about. Not a hobby — a serious professional field. Behavioral economics, systems thinking, design, experimental method, history of a specific industry, anthropology, whatever.

Commit this week to spending three hours a week, every week, for the next year, consuming material from that field. Books. Papers. Practitioners.

One year from now, you will have one new arm on the starfish. Two years from now, two. By the time you are five years into this practice, your career is unrecognizable from where you started, in a good way. You will be the person in your field who has a weird vocabulary nobody else has. That weird vocabulary is where your next decade of income lives.

— — —

Science Anchor

- Frans Johansson, *The Medici Effect* (2004). Still the clearest articulation of why combination beats specialization in innovation contexts.
- David Epstein, *Range: Why Generalists Triumph in a Specialized World* (2019). The modern sequel to Johansson. Also easier to read.

- Research on patent combinatorial novelty — specifically the work showing that high-impact patents tend to combine prior art from previously unconnected domains, not to advance deep within a single domain.
- Margaret Boden's philosophical work on the nature of creativity. Her taxonomy of creativity — combinatorial, exploratory, transformational — is useful, and helps you see that combinatorial is both the most common type and the most teachable.

— — —

Close

Jamal is in a client meeting on a Tuesday morning. The client, a mid-market direct-to-consumer brand, is describing a conversion rate problem on their checkout page. Jamal listens. He asks three questions.

Within six minutes he has diagnosed that the problem is not design, and not copy, and not pricing. It is a behavioral economics problem — specifically, a loss-aversion frame that is making customers anchor on what they are giving up rather than what they are getting. He knows this because he read a paper about loss aversion in 2019 that had nothing to do with e-commerce, and because he has tested variations of loss-aversion framing across seventeen unrelated clients over the last five years, and because he designed an experimental protocol two years ago specifically to isolate this kind of effect.

Nobody else at the meeting has the five disciplines in their head to make that diagnosis. His competitors at other agencies cannot make it. His client's in-house team cannot make it. No AI agent on earth, in 2026, can make it, because the training data does not explicitly connect the paper, the e-commerce context, and the experimental protocol in a single inference.

Jamal can make it, because he built his head on purpose to make exactly these connections.

You can too. It takes years. It starts this week.

Next chapter: why the Surgeon General's loneliness advisory is the most important business memo of the decade, and why empathy is the only product that cannot be mass-produced.

Chapter 9: *The EQ Economy*

In 2023, the United States Surgeon General — the government's top doctor, not an activist, not a pundit, not a podcaster — issued a public health advisory declaring loneliness an epidemic. Not a metaphor. A measurable, quantifiable, shortening-people's-lives epidemic. His data said chronic social isolation raised mortality risk by roughly the same amount as smoking fifteen cigarettes a day. Fifteen. A day.

A decade ago that would have been a national headline. In 2023 it was a Tuesday. Most of us noticed for three minutes and went back to doom-scrolling.

The market noticed longer. The market is still noticing. Because the market saw what a lot of us are too polite to say: a rising tide of lonely, anxious, disconnected humans is not just a tragedy — it is a generational economic opportunity. And the only thing that reliably treats it is another human being, in a room, with their phone face down, giving a damn.

Welcome to the EQ Economy.

— — —

The Loneliness Numbers

Let me lay out the data, because some of you are already skeptical, and you should be. Anyone who uses the word "loneliness" in a business context is sixty percent of the way to selling you a crypto coin.

So here is the data, without the pitch.

Murthy's 2023 advisory compiled evidence from decades of social science research and public health surveillance. The findings were consistent across methodologies: Americans are significantly lonelier, more isolated, and less socially connected than at any comparable point in the modern era. Adults report fewer close friendships than their parents' generation did. Young adults report dramatically fewer. Time spent in face-to-face social interaction has dropped by measurable double-digit percentages since the mid-2000s. The share of Americans who report feeling isolated from others on a regular basis has climbed steadily for two decades and jumped sharply during the 2020 pandemic, then failed to fully recover.

The downstream health effects are not speculative. Chronic loneliness correlates with increased rates of cardiovascular disease, dementia, depression, anxiety, substance abuse, and premature mortality. It is not a mood problem. It is a public health problem with a biological mechanism — prolonged activation of the stress response, elevated inflammation markers, compromised immune function — and it is hitting every demographic cohort in America, though young adults and older adults are getting hit hardest.

What this means in economic terms is that demand for genuine human connection — therapeutic, medical, social, professional — is rising faster than supply can match. And the thing AI reliably cannot deliver, no matter how well it writes sonnets or diagnoses rashes, is *a person who is actually in the room with you, giving you attention.*

The EQ Economy is the set of professions that deliver on that last specification. Those professions are having a moment that is not a moment. It is a structural, demographic, multi-decade shift.

— — —

Why AI Companionship Hits a Ceiling

You might be wondering whether AI is eventually going to solve this. Spoiler: no, and the reason is interesting.

A number of companies have attempted to build AI companion products — apps designed to simulate friendship, romantic attention, therapeutic presence. The early adoption curves looked promising. Users logged in. They had emotional reactions. Some of them reported feeling real comfort. It looked, briefly, like a market was about to form around replacing human connection with large-language-model simulations of it.

Then the retention curves came in, and they told a different story.

Users of AI companionship products, on average, exhibit a characteristic pattern: strong initial engagement, followed by a slow erosion of emotional investment, followed by abandonment. The reasons, pieced together from user interviews and app analytics, cluster around a single insight. Users eventually notice that the AI is not real. Not intellectually notice — they knew that going in. *Emotionally* notice. The connection starts to feel hollow in a way they cannot always articulate but consistently describe.

The uncanny valley of intimacy is real. It is, if anything, more severe than the uncanny valley of physical appearance. A robot that looks almost human feels creepy; a conversation that almost feels intimate feels *sad.* Users report ending their AI companionships not because the AI got worse, but because they became increasingly aware that they were being met by a mirror, not a person. The thing that humans need from other humans — attention from an entity that could choose not to give it — cannot be simulated by an entity that has no choice in the matter.

This is not an anti-AI screed. I use AI tools every day. I am writing a book partly about how to use them well. But the specific economic niche of "person who actually cares about you and is present with you" is one of the few niches that is structurally protected from AI substitution. Probably permanently. Certainly for the next twenty years.

Which means the humans who can deliver that — skillfully, reliably, with real training — are going to be very, very valuable.

— — —

The Presence Premium

"Presence Premium" is my shorthand for the wage uplift that attaches to jobs in which the main economic value delivered is human presence, human attention, and human empathy.

Fields riding the Presence Premium in 2026 include, at minimum:

- **Therapy and mental health counseling.** Demand is outpacing supply at a rate that would be comical if it were not so tragic. Waitlists of six-to-eighteen months are becoming normal in most major cities.

- **Coaching.** Executive coaching, life coaching, career coaching — the credentialed, evidence-based versions, not the Instagram variety. A working coach with real training can charge two to five hundred dollars an hour and be fully booked.

- **Teaching, tutoring, and education.** Especially the one-to-one and small-group versions. AI tutors exist and are useful; they do not replace the human who notices that a specific kid is having a bad week and adjusts the lesson accordingly.
- **High-touch sales.** Not commodity sales, which AI is eating. Relationship-based enterprise sales, complex B2B sales, luxury sales — any sales role where the trust between buyer and seller is the actual product.
- **Caregiving and direct patient care.** Nurses, physical therapists, hospice workers, eldercare professionals. These jobs have a physicality premium *and* a presence premium stacked on top of each other. We will dedicate a full chapter to them in Part Four.
- **Hospitality at the high end.** The maître d' at a Michelin-star restaurant. The concierge at a five-star hotel. The head sommelier. Not the commodity service roles, which are getting automated, but the judgment-plus-presence roles at the top of the ladder.
- **Clergy and spiritual direction.** Secular culture keeps predicting the decline of religious professions. Actual demand for trained humans who sit with people in hard times and provide spiritual presence is holding steady or rising.

Notice the pattern. Every one of these jobs involves a human in a room with another human, giving genuine attention, making adjustments in real time based on what the other person needs, and absorbing some emotional load that the other person cannot carry alone.

Every one of these jobs is getting more valuable, not less, as the rest of the economy automates around them.

— — —

Meet Dr. Eleanor Voss

Dr. Eleanor Voss is sixty-three years old. She is a psychiatrist in Boston, in private practice. Her practice is closed to new patients. Her waitlist — the list of people who have specifically asked to see her, ranked by referral urgency — is currently eighteen months long, and she has stopped accepting new additions to it because she feels guilty about the wait.

Eleanor trained in the 1990s, when the prevailing wisdom in medicine was that psychiatry was a second-tier specialty, that the future of mental health was pharmacology, and that anyone with talent should go into a surgical subspecialty instead. She chose psychiatry anyway, because she liked talking to people, and because her father, who was a factory supervisor, had died of cardiac disease at sixty-four after a decade of quiet depression nobody in the family had known to name.

Thirty years later, Eleanor is among the highest-income practitioners in her specialty in Massachusetts. Her practice is entirely cash-pay — she dropped insurance panels in 2015 because the paperwork was eating her life — and she charges rates that fill the book anyway, because her reputation has followed her from her teaching years at Mass General. She sees twenty-two patients a week. She does supervision and consultation for a half-dozen younger clinicians. She writes a newsletter. She has been interviewed on four podcasts this year.

She has no intention of retiring. Why would she? She is doing the most meaningful work of her life, at rates that let her fund her grandchildren's educations, in a field where the demand is so far beyond the supply that every competent practitioner is going to be busy for the next twenty years.

When I asked Eleanor about AI therapy apps, she thought for a long moment before answering. Then she said, quietly: "They might help somebody get through a hard night. I am glad they exist. But nobody has ever cried to an app and felt afterward that the app was changed by the crying. When my patient cries in this office, I am changed. That is the actual medicine. The AI cannot do that. I do not think it ever will."

Eleanor is not anxious about her career. Eleanor is the career. She built it in a field that had a Presence Premium before anyone called it that, and the premium is compounding as she ages, not depreciating.

— — —

EQ Is Trainable

I want to head off a common objection: "I am not a people person." Great. Neither were most of the people now making a living in the EQ Economy when they started out.

Emotional intelligence, like taste and adaptability, is a trainable skill, not a fixed personality trait. Decades of research on therapist effectiveness, teacher effectiveness, nurse effectiveness, and sales effectiveness converge on a short list of specific behaviors that make humans better at being present for other humans:

- **Active listening.** Genuinely attending to what the other person is saying, reflecting it back accurately, asking questions that show you heard what they meant, not just what they said.
- **Holding silence.** Not rushing to fill the gap. Not inserting yourself. Letting the other person continue, at their own pace, to where they actually needed to go.
- **Naming emotion without judgment.** "You sound frustrated." Not "you should not be frustrated." Not "I understand." Not "I know how you feel." Naming, flatly, what you are noticing.
- **Withholding unsolicited advice.** Nine times out of ten, the person does not want your fix. They want to be heard. The advice only lands if they ask.
- **Attuning to nonverbal cues.** Most emotional information comes through tone, pace, posture, not word choice. The trained professional tracks all of them.

Notice that none of these behaviors require you to be naturally warm, extroverted, or emotionally expressive. They require discipline and practice. Plenty of the best therapists I have met are, in private, fairly reserved introverts. They just put the reps in.

The "three questions before any answer" rule is a good starter practice. Whenever someone brings you a problem — at work, at home, anywhere — commit to asking three clarifying, non-leading questions before you offer a solution. Just that. It sounds absurdly simple. In practice, it rewires how people experience being around you within a few weeks.

— — —

Diagnostic

- Do people seek you out when they are having a bad day? Or do they avoid you, because they know you will try to fix them?
- In a one-hour conversation with a colleague about something difficult, what percentage of the talking time is theirs, and what percentage is yours?
- Can you name the last three times somebody confided something genuinely hard in you? If you cannot, people are reading a signal off you that says "not safe to open up to this person."
- When someone brings you a problem, is your first move to ask questions, or to offer solutions?
- Do you feel drained after emotionally heavy conversations, or energized? Neither answer is wrong, but the answer tells you something about which role in the EQ Economy fits you.

— — —

Monday Morning

Pick one relationship at work this week — a colleague, a direct report, a client — and try the three-questions rule in every conversation for a week. Track what changes. Write down what you notice.

If it works — if the relationship visibly improves, if the person opens up more, if you leave the interaction with a clearer sense of what they actually needed — consider what it would mean to run your whole professional life on this operating system. The answer is probably: it would mean a career pivot. Which might be exactly what this chapter is nudging you toward.

If you have ever been told, by more than one person, that you are a "good listener" — take that feedback seriously. That is the market telling you where you fit. Most of us ignore that signal and go do something career-safe that makes us miserable. Consider ignoring the career advice and following the signal instead.

— — —

Science Anchor

- Vivek Murthy, *Our Epidemic of Loneliness and Isolation: The U.S. Surgeon General's Advisory on the Healing Effects of Social Connection and Community* (2023). Free, public, and remarkably readable for a government document.
- Robert Waldinger's Harvard Study of Adult Development. The longest-running study of human happiness in history. The punchline — that the single strongest predictor of life satisfaction is the quality of close relationships — is not news, but the data behind it has only gotten stronger.
- Research on AI companion retention, particularly studies tracking long-term engagement and emotional outcomes in products like Replika and Character.ai. The consistent finding: initial engagement does not translate to durable emotional benefit.

- Carl Rogers on the therapeutic conditions. Old research, still the foundation. The three conditions of genuineness, unconditional positive regard, and empathic understanding remain the clearest description of what presence actually is.

— — —

Close

Eleanor finishes her last session at 6:47 p.m. on a Thursday. She writes her notes. She walks down Commonwealth Avenue in the spring evening. She is not tired in the way she used to be tired at thirty-five. She is tired in the way a stonemason is tired at the end of a long day of work they meant to be doing. It feels, on balance, good.

Her next patient is scheduled for Monday at 8 a.m. After that, Tuesday at 9, Tuesday at 11, Tuesday at 2. Every slot in her book is spoken for through next April. There is not a recession in sight for her. There is no automation threat on the horizon. Her market is growing faster than the entire profession can train new practitioners, and the profession is itself failing to scale. Her career is a secular tailwind.

If you have ever wondered what a genuinely Presence-Premium career looks like from the inside: this is it. It is boring. It is slow to build. It is decades of reps. It is also, at this point in history, one of the most resilient career arcs a human being can choose.

Pay attention to what people tell you about your listening. If they are telling you something specific, consider believing them.

Next chapter: the new literacy. Not coding. Not prompt engineering. Orchestration. How to build systems, not just send messages.

PART THREE

The New Literacy

The baseline has changed. Here is what every knowledge worker now has to know.

Chapter 10: *Prompt Architecture*

Typing "write me a blog post" into ChatGPT is not a skill. That is just being conscious in 2026.

Let me tell you about two paralegals.

They work at a mid-sized Chicago law firm. Same title. Same pay grade. Same age. Let us call them Amy and Becca.

Amy uses ChatGPT. Specifically, she uses it to summarize contracts. She copies a contract into the chat window. She types "summarize this." She reads the summary. She passes it to the attorney. This saves her maybe twenty minutes per contract. Amy considers herself "good at AI." She mentions it on her LinkedIn.

Becca has a system.

Becca has six Claude Projects running, each tuned for a different contract type. She has a workflow that pulls new contracts from the firm's document management system, runs them through a first-pass summarization agent tuned to her firm's specific practice areas, routes suspect clauses to a second agent that flags them against a database of risky provisions, pushes the resulting flagged summaries into a shared Notion page, and pings the supervising attorney via Slack when anything requires human judgment.

Becca's system runs while she is asleep. Becca's system has completed the morning triage by the time she gets her coffee. Becca spends her workday on the ten percent of documents that actually need a human paralegal — the exceptions, the weird cases, the ones her system flagged for attention.

Amy does roughly three contracts a day. Becca's system does about forty. Becca reviews five of them closely.

Becca is going to get promoted. Amy is going to get laid off. Neither of them is the villain in this story. The difference between them is not intelligence. It is not work ethic. It is not seniority. It is literacy.

Welcome to the new literacy.

———

Prompt, Workflow, System

Fluency with AI tools has three levels, and most American knowledge workers are stuck at level one.

Level one: Prompt. You type a request into a chat window. You read the answer. You use the answer. This is useful. It is also what someone could have done in 2023 with basically no training. It is table stakes. It is not a career.

Level two: Workflow. You chain multiple AI interactions together to accomplish a specific repeatable task. You have a prompt that takes a raw input, transforms it, passes it to another prompt that refines the output, passes that to a third prompt that formats it for delivery. You have created a small production line inside an AI system. The same task that took forty minutes at level one takes eight minutes at level two, because you do not have to think about the individual prompts anymore. You have templated the process.

Level three: System. You have orchestrated multiple AI agents, integrated with other tools, running partially or fully autonomously. Your system pulls inputs from somewhere, processes them through a chain of specialized agents, writes outputs to somewhere, and alerts you when human judgment is required. Your system is running right now, while you are reading this book. You are a supervisor of AI labor, not a user of AI tools. This is what Becca has. This is what Jamal from Chapter Eight has, at scale.

Most of America is at level one. A small minority has crossed into level two. A tiny fraction has reached level three. The wage delta between these levels is enormous and widening.

If you are reading this book and you are still at level one, that is fine. Level one is where everyone starts. But you need to have a plan for moving to level two within ninety days, and to level three within a year. The plan is the rest of this chapter.

— — —

A Concrete Five-Agent Workflow

Let me walk you through a real workflow. I will keep it simple on purpose, because the point is not the specific task — it is the pattern.

Let us say your job involves writing a lot of research briefs. Each brief requires reading a client's materials, pulling in relevant external research, synthesizing it, drafting, and formatting. Under the old workflow, this is a full day of human work per brief. Under a level-three workflow, it is roughly forty minutes of your time per brief, with the rest handled by your agents.

Here is the architecture:

Agent one: The Ingestor. Takes the client's raw materials — emails, documents, notes, transcripts — and produces a structured summary of what the client actually cares about, what they have said, what they have not said, and what the implicit open questions are. Tuned specifically for your industry's jargon. Runs first, independently.

Agent two: The Researcher. Takes the open questions from Agent One and does external research. Finds relevant studies, market data, competitor information. Produces a structured research memo with sources and confidence levels. Runs in parallel or after Agent One.

Agent three: The Synthesizer. Takes outputs from Agents One and Two and produces a draft synthesis — where the client's situation intersects with the external research, what the non-obvious connections are, what the strategic implications might be. This is the conceptual heavy lifting.

Agent four: The Drafter. Takes the synthesis and produces a written brief in your house style. Respects your firm's voice, formatting conventions, and length norms. Produces a clean, readable, client-facing draft.

Agent five: The Checker. Takes the draft and reviews it against a checklist you wrote — does every claim have a source, are there any assertions that need a caveat, is the tone appropriate for the specific client, are there legal or compliance flags to surface before the brief goes out the door.

You sit on top. Your job is to review the final output, apply the judgment that your agents cannot, add the one or two insights that require you to know this client personally, and sign your name to the brief.

Forty minutes of your time. The agents do the rest. This is not a fantasy. This is a workflow you could build, with commercially available tools, in a weekend.

If you could, why would you not?

— — —

Why "Prompt Engineer" Is Not a Career

A couple of years ago, "prompt engineer" was briefly a trending job title. You could find LinkedIn posts where people claimed to be making two hundred thousand dollars a year writing prompts. Some of those posts were real. Most were marketing for prompt-writing courses.

Here is the truth. Prompt engineering, as a standalone skill, is dying. Not because prompts do not matter — they still do — but because writing an effective single prompt is a baseline competence, not a specialization. It is the 2026 equivalent of being "really good at Excel formulas" in 1998. Valuable, yes. A career, no.

The career is in *workflow and system design.* In knowing how to decompose a business problem into a sequence of agent tasks. In knowing when to use one agent and when to use five. In knowing how to route outputs, handle failures, insert human checkpoints, and monitor performance. In knowing which parts of the work genuinely require the full power of a frontier model and which parts can run on something cheaper. In knowing how to make the whole system reliable enough to let run without constant supervision.

These are engineering problems in the broad sense — problems of design and architecture, not problems of typing the right words into a box. And like all engineering problems, the practitioners who can solve them at the system level are going to be paid very well, while the practitioners who can only produce individual outputs on demand are going to see their wages collapse.

If you want a one-line takeaway: stop thinking about what you can make an AI *do* in a single conversation. Start thinking about what you can make an AI *run*, reliably, without you, for days at a time.

— — —

Tool Use vs. Tool Mastery

I draw a distinction in this book between tool use and tool mastery. It is the difference between Amy and Becca.

Tool use is what most people do with AI. You have a specific, immediate problem. You reach for the tool. You solve the problem. You put the tool down. The AI is passive until you summon it. It is a very nice calculator.

Tool mastery is what Becca does. You design systems in which the AI is continuously active, performing specialized tasks inside a broader workflow you control. The AI is ambient. It is not something you reach for; it is something that is already running, in the background, handling work that would otherwise be yours. Your job shifts from executing tasks to orchestrating a small fleet of non-human workers.

The wage gap between tool use and tool mastery is already large and will only grow. Tool users can do, optimistically, one-point-five to two times as much work as someone without AI. Tool masters can do ten to thirty times as much work, depending on the field, while spending their own cognition on the parts that matter.

Not everyone needs to become a tool master. But if you are in any field that involves significant amounts of describable, repeatable knowledge work, and you are not at least climbing toward tool mastery, your career is going to compress. Fast.

— — —

Diagnostic

- In the last month, how many times did I use AI as a one-off tool? How many times did I build or modify a persistent workflow?
- Do I have any AI-driven system that runs without me actively prompting it? Or is every interaction still a manual one?
- Can I articulate, in three sentences, what "agent orchestration" means? If not, that is the next thing I need to learn.
- When I look at a recurring task in my work, do I instinctively think "how do I get this to run without me?" Or do I still think "how do I get better at doing this?"
- Am I using tools that integrate with each other, or am I copying and pasting between tools that do not know about each other?

— — —

Monday Morning

Pick one recurring task in your work this week. Any task that you do more than once a week. Any task that has a predictable shape, takes more than thirty minutes, and is mostly information work.

This month, build a multi-agent workflow for it. Not a single prompt. A workflow. Minimum three agents handing off to each other. It does not have to be perfect. It has to exist.

Then, ship it. Use it weekly. Refine it. Add a fourth agent. Then a fifth. Within ninety days, that workflow should be running reliably enough that you are checking its outputs rather than executing the task yourself.

Next quarter, pick another task. Build another workflow. Then another. Within a year, a material fraction of your current job is running on systems you built. Within two years, those systems are the biggest asset on your career's balance sheet.

Do not wait for your employer to tell you to do this. Your employer is slower than the market. By the time they formalize "AI literacy training," the people who have been doing this on their own for two years will already be the ones promoted into the roles designing that training.

— — —

Science Anchor

- Anthropic's agent research and documentation, which is public and accessible. Start with their writing on tool use and agent design patterns.
- OpenAI and Google DeepMind research papers on multi-agent systems. Not required reading, but useful if you want to understand why the frontier is moving where it is moving.
- Ethan Mollick's ongoing writing on "centaur" and "cyborg" AI use patterns. The centaur pattern maps closely onto

workflow-level fluency. The cyborg pattern maps onto system-level fluency.

- The MCP (Model Context Protocol) documentation, which describes how agents interoperate with external tools. This is becoming the plumbing of system-level AI work.

— — —

Close

Amy is told, in Q3, that her role is being "restructured." She is offered a severance package. She takes it because she has no leverage. She updates her LinkedIn. Her next job, if she finds one in the next six months, will be at a lower salary than this one.

Becca is promoted in the same quarter. Her firm asks her to teach a two-day workshop to the rest of the paralegal team. She demurs — she is busy — but agrees to a reduced version. The partner who requested the workshop quietly doubles her salary a month later, because the firm has realized that she is doing the work of four paralegals and they would rather pay her twice what she used to make than lose her.

Neither of them did anything morally superior or inferior. They made different literacy decisions in 2024 and 2025. Becca decided to learn how to orchestrate AI systems. Amy decided to learn how to use AI tools. That difference, compounded over two years, produced dramatically different outcomes.

You are standing at the same fork right now. The good news is the fork is still open. The bad news is it will not stay open much longer. In another two to three years, the people who crossed over to system-level fluency will be the only ones still getting hired for knowledge-work roles, and the rest will be absorbed into whatever jobs are left that do not require this skill.

The new literacy is not coding. It is orchestration. Start orchestrating this week.

Next chapter: your agent fleet. Because by 2030, every professional is going to have more direct reports than most middle managers — and those reports are going to be algorithms.

Chapter 11: *Agent Wrangling*

You have direct reports now.

You did not hire them. You did not interview them. You did not negotiate their salaries. They do not go to HR when they are unhappy, because they are never unhappy, and because they have no concept of HR. They work nights, weekends, and holidays. They never ask for a raise. They never push back on a deadline. They also, occasionally, fabricate entire facts with such confidence that you will nod along for eleven minutes before you notice something is wrong.

Congratulations. You are a middle manager now. Of algorithms.

This is not a metaphor. This is the job description for a large and growing share of knowledge work in 2026. If you are building anything on top of AI systems more sophisticated than a chatbot — and by 2028 most people will be — you are managing non-human workers. You need to be good at it. Most people are, at the moment, extremely not good at it.

Welcome to agent wrangling.

— — —

What an Agent Actually Is

Let me give you the working definition, because the word "agent" is being used three different ways by three different industries and it is muddling the conversation.

An AI agent, in the sense that matters for your career, is a large language model hooked up to tools and given a goal. The model can make decisions, take actions, use external systems, and operate with some degree of autonomy between the moments you check on it. It is not just a chatbot you prompt and listen to. It is a worker you delegate to.

The distinction matters because delegation and prompting are fundamentally different skills.

Prompting is what you do with a chatbot. You type a request. It responds. You decide what to do with the response. You are in the loop for every cycle.

Delegation is what you do with an agent. You describe the goal. You describe the constraints. You grant authority to take certain actions within those constraints. Then you walk away. When you come back — an hour later, a day later, a week later — the agent has either completed the task, surfaced a decision you need to make, or failed in some specific way you need to understand.

Most people in 2026 are still prompting. A much smaller group is delegating. The wage gap between the two groups is growing so fast that by 2028 it will be the main career distinction in most knowledge-work fields.

— — —

Your New Middle Management

Here is the structural irony nobody wants to name out loud.

Middle management as a human career category is collapsing. Most of what a middle manager traditionally did — schedule work, route tasks, monitor progress, handle exceptions, escalate decisions — can now be done by software. Org charts are getting flatter. Ranks of middle managers are getting thinner. A lot of the people reading this book are middle managers who are trying to figure out what is left for them.

What is left is this: someone has to do all of those same middle-management functions for the *agents*. Someone has to scope their work, route tasks to the right agent, monitor outputs, handle exceptions, and escalate decisions. Someone has to be the boss of the bots.

This is the career that is growing. Not "AI engineer," which is a specific technical role. Not "prompt engineer," which we established last chapter is dying. *Agent manager.* The person who designs, supervises, and accepts responsibility for a small or large fleet of AI workers.

The skills transfer directly from human middle management, which is good news for the millions of middle managers being displaced. You already know how to scope a task, set expectations, monitor outputs, and catch a team member drifting off course. You just have to apply those skills to direct reports who happen to be software.

The parts that do not transfer are the parts everyone forgets. Agents fail differently from humans. Agents do not have institutional loyalty. Agents do not pick up tacit context the way a new hire does in their first three weeks. Agents drift. You have to know how to wrangle them specifically.

— — —

How Agents Fail

If you have ever managed humans, you know the failure modes. Missed deadlines. Low quality work. Interpersonal conflict. Burnout. Dishonesty. Territorial nonsense. Getting sick. Getting bored.

Agents have none of those failure modes. They have their own, and they are worse in some ways because they are unfamiliar.

Hallucination. An agent can produce confidently false output. It can cite sources that do not exist, quote statistics it made up, and reason from premises that are factually wrong. It will never tell you it is doing this, because it does not know it is doing it. A human who made up sources would be humiliated when caught. An agent will do it again on the next task and sound just as confident.

Scope drift. An agent asked to do task A will sometimes do tasks A, B, and C, because B and C seemed adjacent and helpful. This is charming when it works. It is a compliance nightmare when the agent, trying to be helpful, takes an action you specifically did not authorize.

Context loss. An agent operating over a long task will forget or misremember earlier parts of its own work. Unlike a human, it does not notice this happening. It just writes past the contradiction and moves on.

Silent drift. An agent running on a schedule, over weeks or months, will slowly produce outputs that diverge from what it was producing at launch — because the inputs change, the model changes, or the scaffolding around the agent changes. Nobody notices until something specific breaks.

Over-confidence. Agents, by default, are calibrated poorly. They express certainty about uncertain things. They fail to flag when they are out of their depth. This is a management problem, not a model problem, and it is solved by prompt discipline and observability — not by waiting for the next model release.

These are engineering problems. They are also management problems. A good agent wrangler treats them the way a good middle manager treats team dysfunction: as things you catch by paying attention, not things you pretend will sort themselves out.

———

The Five Principles of Agent Management

After two years of building and running agent systems, I have converged on five principles that work across domains. Treat these as a starter template. Refine them for your specific context.

- **Clear Scope.** Every agent should have a single-sentence description of what it is authorized to do. Not a paragraph. A sentence. If you cannot describe the agent's job in a sentence, the scope is wrong. Re-scope before you deploy.
- **Bounded Authority.** Every agent should have an explicit list of what it can do without asking permission and what requires human sign-off. The list should be tight at first and loosen only as the agent proves itself. "Send email" and "book meeting" are different authority levels. "Spend money" is its own category. Think like a manager delegating to a new hire.
- **Observable Outputs.** You should be able to see what the agent is doing, when, with what inputs, producing what outputs. Logs, traces, sample reviews. If you cannot audit it, you cannot manage it. If you cannot manage it, you

cannot be responsible for it. If you cannot be responsible for it, you cannot deploy it.

- **Fail-Safe Defaults.** When the agent is confused, what does it do? It should halt, flag, or escalate — never improvise. Agents improvising is where the horror stories come from. A good agent, when uncertain, acts like a junior analyst on their first day: asks rather than guesses.
- **Escalation Paths.** Certain triggers should always pull a human into the loop. Large financial commitments. Sensitive customer interactions. Legal or compliance flags. Unusual patterns that the agent itself cannot interpret. Design these triggers into the system, not bolted on after an incident.

Do this work up front, before you deploy. It is ten times cheaper than trying to put it in after the agent has been running unsupervised for three months and generated a forty-page incident report.

— — —

Diagnostic

13. Do I currently have any AI agents running on my behalf? How many?
14. Could I describe the scope of each agent in one sentence?
15. Do I know what each agent is authorized to do without asking me?
16. When was the last time I reviewed a sample of an agent's outputs to check for drift?

17. If an agent made a significant error tomorrow, would I know about it before the client, the regulator, or the Monday meeting?

— — —

Monday Morning

This week, build one agent that does something real, runs on a schedule, and reports to you.

It does not have to be complicated. It can be a simple workflow that reviews your inbox overnight and produces a prioritized summary. It can be a research assistant that finds and summarizes three articles in your field daily. It can be a formatter that converts your raw notes into client-facing drafts.

The point is not the specific task. The point is the practice. You design the scope. You set the authority. You build the observability. You define the escalation. You deploy it. You monitor it for a week. You refine it.

At the end of the week, you will have personally experienced every failure mode in this chapter. Each one is a lesson you cannot get from reading. Once you have wrangled one agent for a week, the concepts are yours for life.

— — —

Science Anchor

18. Anthropic's public documentation on tool use, agent architecture, and Constitutional AI. Start there, because it is the clearest practitioner writing available right now.

19. Google DeepMind and OpenAI research on multi-agent coordination and reliability. Skim the abstracts; the technical depth is optional.
20. Reliability engineering literature from the software world — specifically the body of work on site reliability engineering (SRE) and on observability. The principles of managing unreliable software systems transfer directly to managing unreliable agent systems.
21. Classic management literature on delegation — Drucker, for example. The principles of delegating to humans are roughly ninety percent of the principles of delegating to agents. The remaining ten percent is what this chapter is about.

— — —

Close

There is a strange grief in this chapter that I want to name before we close it out.

A lot of the middle managers who are losing their jobs in 2026 were actually good at their jobs. They knew their teams. They knew the business. They knew which escalations were real and which were noise. They were the load-bearing wall of their organizations in ways the C-suite often did not appreciate until after they were gone.

Those same skills — scoping, delegating, monitoring, escalating — are exactly the skills required to manage agents. If you are one of the displaced middle managers, you are not obsolete. You are early. Your specific skill set is about to become extremely valuable, applied to a new kind of worker.

You just have to be willing to manage a team that does not drink coffee and never goes home.

I know people who are thriving in this new role. They are running teams of fifty agents with two or three human collaborators. They are reporting to a CEO rather than to a VP. They are, functionally, running an entire department with a headcount of one plus their fleet. Their comp has not decreased. It has increased.

This is not a fantasy. This is a career path. It is the career path for a lot of the people who are reading this book with knots in their stomachs about where they go next.

You go here. You become a wrangler.

Next chapter: why the most unsexy, boring, eye-glaze-inducing career in 2026 is also the most insulated from the cortex revolution — and why the phrase "compliance officer" is about to become code for "absurdly well-paid."

Chapter 12: *Compliance Native*

The sexiest career in 2030 is GDPR specialist. I am dead serious. Stop laughing.

I know how that sentence reads. I know what "compliance" sounds like to most knowledge workers. It sounds like the thing your HR department makes you watch a 47-minute video about every January. It sounds like middle-aged people in blazers explaining why you cannot send that spreadsheet to that person. It sounds like the field you go into when you have given up.

It is not. It is about to be the most insulated career in American knowledge work, and the people who recognize this early are going to have an absurd decade.

Let me walk you through why.

— — —

The Invisible Layer

Every AI deployment a company makes creates regulatory exposure. Every single one. Even the boring ones. Especially the boring ones.

When your marketing team licenses an AI tool that touches customer emails, they have just created a GDPR exposure, a CAN-SPAM exposure, and depending on the state, a handful of state-specific privacy exposures. When your HR team pilots an AI tool that scans resumes, they have created exposures under Title VII, the ADA, the EEOC, New York's automated employment decision tool law, Illinois's AI Video Interview Act, and, if they operate in

Europe, the full weight of the EU AI Act's high-risk employment provisions. When your finance team uses an AI to summarize contracts, they have created exposures under SOX, and if the contracts touch healthcare, HIPAA. The list goes on and gets worse.

Most companies deploying AI in 2026 do not have a clear inventory of what they have deployed, what each deployment touches, what each deployment's regulatory footprint is, or who would be responsible if one of them caused a reportable incident. They do not have this inventory because nobody on the team knows enough to build it.

This is the Invisible Layer. It is the regulatory, legal, and governance substrate underneath every AI deployment, and it is the part that nobody is paying attention to until a regulator, an auditor, or a plaintiff's attorney points a finger at it.

If you are the person who can see, map, and manage this layer, you are about to be very, very busy.

———

The Regulatory Stack Is Getting Taller

A brief tour of the landscape, because it changes fast and most people have not looked in a while.

22. **GDPR** (EU, since 2018). Requires specific disclosures and consent for most personal data processing. Penalties up to 4% of global revenue. Extraterritorial — applies to any company with EU data subjects.
23. **The EU AI Act** (rolling out 2024–2026). Classifies AI by risk level. Bans some uses. Heavily regulates "high-risk" uses. Requires human oversight, documentation,

conformity assessment, and registration. Affects any company with EU customers.

24. **HIPAA** (US healthcare). Any AI touching protected health information requires business associate agreements, auditable access controls, and breach notification procedures. Penalties per violation, stacked.
25. **SOX** (US public companies). Financial controls, audit trails, documentation. AI in the finance function pulls AI into SOX scope.
26. **PCI DSS** (payment card data). AI systems touching card data have to meet specific technical and organizational standards.
27. **SOC 2** (not a law, but contractually mandatory for B2B SaaS). Covers security, availability, processing integrity, confidentiality, privacy. AI deployments inside SOC 2 scope require documentation, monitoring, and incident response.
28. **State AI laws** (US). Colorado's comprehensive AI law. California's sector-specific rules. New York's employment AI law. Illinois's biometric and video interview laws. A patchwork, and the patchwork is growing quarter over quarter.
29. **Sector-specific regulators.** FDA for medical AI. SEC for financial AI. FTC for deceptive marketing AI. CFPB for consumer credit AI. DOL for employment AI. Each has its own guidance and its own enforcement posture.

This stack is not shrinking. It is growing in every direction at once. Every new AI deployment, in every sector, creates a new application of old law and invites new specific law. The gap between what companies are deploying and what companies can demonstrably comply with is widening every quarter.

The people who close that gap are going to be paid like the gap is worth closing. Because the gap *is* worth closing. A single privacy incident at a mid-market company can cost seven to nine figures. Avoiding that is cheaper than paying for it. Compliance is the cheapest insurance policy most companies can buy, and the companies figured this out years ago — they just cannot find enough people who can actually do it.

— — —

Compliance Is a Signer Role

Remember Chapter Five. The Signer Role is any position in which the core economic function is "licensed human who accepts responsibility for a regulated output." Compliance is a Signer Role. It always was. It is now stacked on top of the Accountability Premium discussed earlier, and the stack is the thing that makes this career bulletproof.

A corporate compliance officer, in most jurisdictions and most industries, is a personally accountable position. When a regulator investigates a violation, they interview the compliance officer personally. When the company is fined, the compliance officer's name is in the filings. When there is a criminal referral — rare, but real — the compliance officer is sometimes on the hook, not the CEO, because the compliance officer is the designated responsible party.

That sounds like a liability. It is, in fact, the source of the premium. You cannot automate the person whose name goes on the compliance attestation to the regulator. You cannot replace them with an AI tool, because the AI tool cannot be fined or prosecuted. You cannot even outsource them easily, because the regulator requires a specific, accountable, reachable human with authority over the function.

Which means compliance officers are, like physicians and licensed attorneys and licensed engineers, part of the small and growing category of knowledge workers whose economic value *goes up* as AI capabilities expand. The more AI a company deploys, the more compliance surface area exists. The more compliance surface area exists, the more valuable the human responsible for managing it becomes.

This is a secular tailwind that most compliance officers themselves have not yet fully priced in. The ones who have are quietly tripling their rates.

— — —

What "Compliance Native" Means

There is an important distinction between compliance *added* and compliance *native*.

Compliance added is what most companies do. You build the thing. You ship the thing. Then, when a regulator or an auditor asks, you try to reconstruct how the thing works and whether it meets the rules. This is expensive, stressful, and error-prone. It produces the worst kind of compliance work — retroactive documentation of decisions nobody remembers making.

Compliance native is what the good companies do, and what the surviving companies will do. Compliance requirements are designed into the system from day one. Data minimization. Access controls. Audit trails. Documentation. Human oversight. Explainability. These are not added later; they are part of the architecture.

The person who can do this — the "compliance-native" professional — is the person who can sit at the table when an AI system is being designed and say, "Before we go further, here is what GDPR requires, here is what the EU AI Act requires, here is what our internal governance requires, here are the five design choices that will make this system compliant by default instead of broken by default." That person is not a blocker. That person is the reason the system ships at all, in markets you want to sell into.

This is the career. It is technical enough to understand what is being built. It is legal enough to understand what the rules say. It is strategic enough to understand how those rules will shape the business. It does not exist in one academic discipline. You build it by deliberately stacking multiple disciplines, like Jamal in Chapter Eight. Regulatory knowledge plus AI literacy plus business judgment. The starfish shape again, with a specific flavor.

— — —

Who Should Consider This

If you fit any of these profiles, compliance-native careers may be the single highest-leverage pivot available to you in 2026.

30. **Lawyers, especially regulatory lawyers.** You already know the rules. Now learn the technology. You will be in the top five percent of your own profession within two years.

31. **Auditors, risk professionals.** You already know how to verify and document. Adding AI literacy opens a career ceiling that would otherwise be closed.

32. **IT and cybersecurity professionals.** You already know security and systems. Adding privacy, regulatory, and governance knowledge converts you into a senior corporate role.

33. **HR and employment professionals.** AI in hiring, AI in performance management, AI in monitoring — all are getting heavily regulated. You are in the middle of the storm whether you realize it or not.

34. **Engineers with curiosity about the business side.** Technical fluency is rare among compliance officers. If you have it and you can talk to lawyers without rolling your eyes, you are extremely hireable.

35. **Mid-career pivoters.** This is one of the few fields in knowledge work where starting at forty-five is not a disadvantage. Experience, judgment, and discretion are the point of the role.

— — —

Diagnostic

36. Do I know, for my own company, which regulations apply to the AI systems we have deployed?

37. If a regulator asked our compliance officer to produce an inventory of our AI deployments and their regulatory footprints, could the compliance officer do it?

38. In the last year, did I learn anything substantive about privacy law, sector-specific regulation, or AI governance?
39. When a new AI tool is proposed at work, am I in the conversation about what it touches and who could be harmed? If not, should I be?
40. Do I know what CIPP/US, CISSP, CIPP/E, or CIPM are? If not, these are the starter credentials worth understanding.

— — —

Monday Morning

Pick one regulatory framework relevant to your industry. One. Not five. Not all of them.

Spend the next ninety days actually learning it. Read the official text, not the summary articles. Take a course if one is available. If it has a credential attached — CIPP/US for US privacy, CIPP/E for European privacy, CISSP for security, CISA for audit, CCEP for general corporate compliance — put the exam on your calendar for nine to twelve months out.

At the end of ninety days, you will know more about that framework than ninety percent of the people at your company. At the end of a year, with a credential, you are qualified for roles that did not previously fit your resume.

This is not glamorous. You will not brag about it at dinner parties. You will, however, be fully employable in a field that pays well and has a structural worker shortage going into the most regulated decade of American corporate history.

— — —

Science Anchor

41. The EU AI Act, full text. Open it. Skim it. Come back to the parts relevant to your industry.
42. NIST AI Risk Management Framework (AI RMF 1.0 and generative AI profile updates). The US baseline for organizational AI governance, and the single best free reading on the subject.
43. IAPP (International Association of Privacy Professionals) materials and certifications. CIPP/US and CIPP/E are the standard credentials in this space.
44. SOC 2 Trust Services Criteria. Boring. Essential. Widely cited in B2B contracts.
45. Your state bar association or professional licensing board's guidance on AI, if you are in a licensed field. These guidances are updating quickly and are often the first place to see where enforcement is headed.

— — —

Close

The compliance officer is, in 2026, the new chief counsel of the AI age. Thirty years ago, every serious American company had a general counsel who sat at the executive table and vetoed the worst ideas before they became lawsuits. Today, every serious company is building the equivalent role for AI risk — and most of them have not staffed it adequately, because they do not know where to find the people.

You can be one of the people. It will take a year of deliberate work. It will pay back for the rest of your career.

The boring professions are where the money is hiding now. Regulation, licensure, accountability, documentation. Everything the tech industry scoffed at for twenty years is suddenly the moat.

The scoffers are not laughing anymore. They are hiring.

Next chapter: the solo operator. Why a twenty-nine-year-old with two employees and forty agents is making more money than most mid-career executives — and why this pattern is about to become the default for ambitious careers.

Chapter 13: *The Solo Operator*

You met Jamal in Chapter Eight. Twenty-nine years old. Austin. Marketing agency. Two employees. Forty agent workflows. Four million dollars in annual revenue.

I want to bring Jamal back, because Jamal is not a novelty. Jamal is the template.

By 2030, a meaningful share of the most successful careers in knowledge work are going to look like Jamal's career — tiny human headcount, large agent fleet, outsized revenue per human, direct ownership of customer relationships. This is not a fringe prediction. This is what the underlying economics of AI-leveraged work make inevitable, as long as nothing dramatic changes at the regulatory or macroeconomic level.

The question this chapter tries to answer is: is this career path for you, and if so, how do you start?

— — —

A Tuesday in Jamal's Life

Let me walk you through one specific Tuesday.

At 6:30 a.m., Jamal's phone buzzes. It is his agent fleet's morning report. During the night, his research agents scanned a dozen industry publications for client-relevant news. His copywriting agents drafted three blog posts for three different clients in three different voices. His analytics agents pulled last week's campaign numbers for every active client and flagged the three campaigns that need human attention today.

Jamal reads the report in bed, on his phone, in about twelve minutes. He makes three judgment calls — kill one campaign that is underperforming, escalate one blog draft that does not quite match the client's voice, approve the other deliverables as drafted.

By 9 a.m., those decisions have been executed by the agent fleet. The blog posts have been formatted, passed through his compliance checks, sent to the clients for review. The campaign kill has been communicated to the ads platform. The flagged draft has been rewritten, reviewed, and re-queued for his morning eyes.

Jamal spends the rest of his morning on the things the agents cannot do. He has a discovery call with a prospective client — a new industry, a new problem, a conversation where pattern recognition and judgment matter more than execution. He has a one-on-one with one of his two human employees. He has a creative session on a brand identity question for an existing client, where the twelve-second diagnosis of loss-aversion framing from Chapter Eight turns into a specific creative strategy that will pull from four of his five disciplines.

By 3 p.m. he has handled, as a human, the kind of work that would require eight senior people at a traditional agency. Because the other work — the production work, the routine work, the high-volume low-stakes work — has been handled by his fleet.

He is home by six. He works out. He eats with his fiancée. He reads a book for an hour. He does not check his email.

He makes more money than ninety-five percent of marketing executives twice his age. He works fewer hours than he did as a junior copywriter. His business margin is about forty percent, because he has almost no overhead.

This is not an anomaly. This is an emerging operating model. The question is how widely it generalizes.

— — —

Why This Was Impossible Five Years Ago

It is worth pausing on why Jamal could not have built this business in 2019. The answer is instructive, because it tells you what specifically has changed.

Five years ago, a one-person agency would have faced a production tax that made scale impossible. Writing a blog post took four hours. Producing a video took a day. Running an analytics report took half an afternoon. Designing a landing page took a week. To serve ten clients at the quality Jamal's serves, you needed a team of roughly a dozen specialists, and their combined salaries and overhead ate most of the revenue.

Five years ago, the infrastructure tax was also brutal. You needed an office, or at least a well-outfitted remote setup. You needed separate tools for each specialty, each with its own license fees. You needed onboarding, benefits, legal, HR, and the full administrative burden of employing people.

Five years ago, the context tax was the invisible killer. Even if you automated some of the production, every client required substantial context to be held and transferred between specialists. Meetings. Briefs. Hand-offs. Reviews. Most of a specialist's time, in any small agency, was spent on coordination rather than on the work itself.

All three taxes have collapsed.

Production is now inexpensive, because agents handle the execution layer. Infrastructure is inexpensive, because remote tools and cloud-first software have become commoditized. Context is cheap, because a single human holding the full client relationship can communicate it to specialized agents faster than that same human could brief a team of humans.

What is left is the part that was always scarce — strategic judgment, relational trust, taste, accountability. Jamal brings those. His fleet does everything else. The combination is something that literally was not economically viable before 2023 and is going to be extremely economically viable for the next decade at least.

— — —

What the Solo Operator Actually Does

If agents are doing most of the production, what does the human solo operator do all day? A reasonable question. The answer lines up suspiciously well with everything we have been building toward in this book.

46. **Customer-facing relationship work.** Discovery calls. Difficult conversations. Trust-building. Negotiation. Conflict resolution. The parts of business where presence matters, judgment matters, and no client wants to be the account being managed by a bot.
47. **Vision and strategy.** What is this business trying to do? Who is it for? What is the next bet? Where is the market moving? These are questions agents can assist with but cannot own. The solo operator is the strategist-in-chief.

48. **Taste and curation.** Reviewing agent outputs. Killing bad drafts. Escalating good ones. Making the five-percent judgment calls that convert a technically-correct agent deliverable into a client-delighting work product.

49. **System design and maintenance.** Building the workflow. Refining it when it breaks. Adding new agents for new tasks. Retiring agents that have been superseded. The solo operator is the architect of their own factory.

50. **Accountability.** Someone has to sign. Someone has to be responsible when it goes wrong. Someone has to make the call when the situation falls outside the agent's scope. That someone is the solo operator, and by definition, that is not a function you can offload.

Notice what is on that list. Presence. Taste. Strategy. System design. Accountability. Those are the three pillars of this book — Physicality-adjacent presence, Accountability, Taste — plus the new literacy of orchestration. The solo operator operating model is the most extreme application of the human premium. It is the pattern where the human does exactly what only humans can do, and no more than that.

That is also why it is hard. It is not forgiving of weaknesses. You cannot hide behind a team. You cannot specialize your way out of parts of the job that do not suit you. You have to do every one of those five functions, well, every week.

———

The Risks That Nobody Talks About

Let me be honest about the failure modes, because every piece of breathless "solopreneur" content online skips over them.

51. **Burnout.** You are the whole company. When you get sick, the company gets sick. When you need a break, the agents keep running but the client-facing work stops. Many solo operators burn out in year three. The ones who do not are, almost without exception, people who have built deliberate rest into their systems from day one.
52. **Single point of failure.** If your systems go down at a bad time, there is no team to pick up the slack. Good solo operators invest heavily in redundancy, monitoring, and disaster-recovery — not as a hobby, but as a survival necessity.
53. **Isolation.** Running a one-person business in a remote setup is, at a basic human level, lonely. There is no lunchroom. There is no casual banter. There is no team to celebrate wins with. Many solo operators underestimate how much this matters until they are two years in and starting to feel strange.
54. **No succession.** What happens to the business when you want to retire, or step back, or sell? There is no team to step up. The value of the business is almost entirely in your head. This is a solvable problem, but most solo operators do not solve it until they are trying to exit, by which point it is expensive.
55. **Infrastructure overhead you did not anticipate.** Benefits. Taxes. Legal entity. Bookkeeping. Insurance. Contracts. None of these disappear just because you are solo; they become your personal job to manage. This alone burns out a substantial fraction of would-be solo operators in year one.

None of these are dealbreakers. All of them are manageable. But they should be priced into your thinking before you leap, not discovered two years in when you have already committed.

— — —

The Solo Operator Inside a Big Company

Here is the move most of this book's readers should actually make, because it captures most of the upside without the hardest downsides.

You do not have to leave your job to be a solo operator. You can be the solo operator *inside* your company.

This is the "intrapreneur" pattern, updated for the agent era. You identify a function at your company that could be run by one person plus a fleet of agents. You propose to run it. You build the agent infrastructure. You deliver the function at a higher output level than the current staffing could achieve. You collect a salary that reflects the leverage, rather than starting a business from scratch.

This pattern works particularly well for mid-career professionals who have deep domain expertise and substantial organizational capital but who cannot easily walk away from income continuity, benefits, or industry relationships. It captures the leverage of solo operation inside a risk-tolerable container.

The specific things that make this work:

56. You pick a function that has clear, measurable outputs, where your success can be demonstrated numerically rather than politically.

57. You build the agent infrastructure on the company's approved tooling, which means compliance and security are someone else's problem, not yours.
58. You document your systems, so you are not the single point of failure for the company (and so you are credit-worthy for the leverage you are creating).
59. You position yourself as the person who designed the function, not just the person currently operating it. This is how you protect your compensation as the function scales.

A lot of the wealth creation in knowledge work over the next decade is going to happen in this exact pattern. Corporate, salaried, agent-leveraged, with roles that look like the old department but headcount that looks like Jamal's.

— — —

Diagnostic

60. Could I describe the specific tasks in my current job that are genuinely uniquely human, versus the ones that are currently done by humans only because AI has not yet reached them?
61. If my company allowed it, could I run my current function with one-quarter the headcount and an agent fleet? What would break?
62. Do I want to run my own business, truly, or am I attracted to the idea of it? There is a huge difference, and the first two years of solo operation will expose which one I actually am.

63. Do I have the personality structure for solo operation? Specifically: comfort with ambiguity, ability to market and sell, tolerance for isolation, discipline around boundaries.
64. If I stayed corporate, could I be the "solo operator inside" — the person designing agent-leveraged functions within an existing organization?

— — —

Monday Morning

Pick the path this week. Not a permanent commitment — a direction.

If you are leaning toward true solo operation, begin the on-ramp. Start a side project that uses agent leverage. Run it for six months. See if the reality of solo work matches the fantasy. Most people discover something surprising in the first ninety days, good or bad.

If you are leaning toward intrapreneur, identify the function in your current company that you could plausibly run at higher leverage. Write a one-page proposal. Keep it in a drawer for now. Refine it monthly. When the right organizational moment arrives — a reorg, a budget crunch, a leadership change — pull it out.

If you are not sure, do both in parallel. The side project and the internal proposal reinforce each other, and your choice will reveal itself.

Do not, under any circumstances, quit your job this week to become a solopreneur because you read an inspiring chapter in a book. That is how bad decisions get made. The best solo operators of 2028 will be people who spent 2026 and 2027 quietly building the foundations while they were still employed.

— — —

Science Anchor

65. US Census Bureau data on small business and nonemployer firm formation. Look at the 2020–2026 trend. Solo-operator firm formation is setting records.
66. Research on optimal firm size in knowledge-heavy industries, particularly the work on coordination costs. The coordination cost savings from very small firms used to be offset by production costs. That tradeoff has changed.
67. Writing by Pieter Levels, Rob Walling, and others on bootstrapped and tiny-team businesses. Not required reading, but useful as case studies.
68. Tim Ferriss's *The 4-Hour Workweek* (2007). Old, uneven, partially dated — but the core insight that lifestyle-first small business design is a valid career path anticipated where a lot of knowledge workers are now headed. Worth skimming once.

— — —

Close

The twentieth-century career looked like this. You joined a firm. You climbed the ladder. You specialized. You got promoted. You retired.

The twenty-first-century career looks like this. You build a fleet. You run it. The fleet is yours — either literally, as Jamal's is, or functionally, as a corporate intrapreneur's is. You do the parts of the work that only humans can do. You take the economics that used to go to five people and keep them.

Not everyone will run this career. It does not fit everyone. It should not. But a meaningful share of the most ambitious, most adaptable, most agent-fluent professionals of the next decade will end up here, either solo or inside a company, either by design or by accident.

If that is you, you do not need to wait for permission. The tools exist. The market is ready. The client demand is already there. Everything you need to begin is in reach this week.

Jamal, by the way, spent 2022 through 2024 working at a traditional agency, quietly building skills and saving runway. He did not quit until he had three paying clients and enough savings for a year. He is not reckless. He is prepared.

Be prepared. Start preparing this week.

Next chapter: we stop talking about where the premium is going and get brutally honest about where it is leaving. Part Four begins with the autopsy of the jobs that are not surviving this transition. The one with your role on the list, if you are unlucky. Let us look at it together.

PART FOUR

The Career Graveyard and Gold Rush

The honest map of what is dying, what is booming, and where to plant your flag.

Chapter 14: *White-Collar Autopsy*

This is the hardest chapter to write in the book. It is also the most important.

For thirteen chapters I have been telling you where the human premium lives, how to build toward it, how to pivot into it. That is the optimistic half of the book, and it is honest, and it is not false. The premium is real. The careers described in the first half are happening.

But there is another half. There is the part where we talk honestly about the jobs that are not going to survive this transition, and the people who are in those jobs right now, and what happens to them. A book that skipped this chapter would be doing a disservice to readers who need straight information and who already suspect the truth. So let us look at it together.

I will not enjoy this chapter. You might not either. But you will trust the rest of the book more if I do not pretend it is all upside.

Welcome to the autopsy.

— — —

Who Is on the Slab

Here is the honest list, compiled from the Goldman Sachs 2023 exposure analysis, the Anthropic Economic Index, McKinsey's 2024 occupational projections, and labor-market data from 2023–2026. These are the white-collar roles with the most severe task exposure to current AI systems, where real employment declines are already visible in the data.

69. **Junior paralegals and legal assistants.** Document review, basic research, first-pass contract summarization. Employment levels flat to declining while firm output per paralegal rises sharply.

70. **Customer service representatives (front-line, scripted).** Chatbots and voice AI are absorbing the high-volume low-complexity tier at most large customer service operations. Headcount reductions of 20% and higher are already documented at specific firms.

71. **Junior analysts (financial, consulting, research).** Summarization, chart-making, basic modeling, deck production. The tasks that used to train a junior analyst into a senior analyst are being done by AI, which is hollowing out the bottom of the career pipeline.

72. **Junior copywriters and content marketers.** Blog posts, email drafts, social media content, basic ad copy. The market for this work is compressing rapidly. What is left is being consolidated into fewer, more senior "AI-leveraged" roles.

73. **Mid-level marketers producing rather than strategizing.** The middle of the marketing career ladder — people who produce deliverables rather than direct strategy — is being squeezed from both ends.

74. **Basic accounting, bookkeeping, and tax preparation.** Software has been eating the low end for twenty years. AI is eating the middle now. What is left is specialist, licensed, accountability-heavy work at the top.

75. **Middle management in knowledge industries.** As discussed in Chapter Eleven, the routing-and-coordinating

function is being increasingly done by software. Middle managers whose primary job was task allocation and status reporting are in the most vulnerable position in the entire white-collar economy.

76. **Junior developers in certain segments.** Specifically: web front-end work, boilerplate back-end work, basic QA, and repetitive feature work. Senior and specialist developers are not on this list. The juniors are, because the rungs of the ladder have been kicked out.

77. **Administrative assistants and schedulers.** A long-running trend, now accelerated. Voice and workflow AI absorbs most of what the role used to do.

78. **Translators and transcriptionists, non-specialized.** General-purpose translation has collapsed in cost. Specialist translation (legal, medical, literary, diplomatic) is holding steady to rising.

Notice a pattern on this list. Most of these roles are not elite, but they are not entry-level either. They are the middle of the middle. The ones where the job is describable in a paragraph, the outputs are structured, the quality bar is fair rather than exceptional, and the training pipeline into senior roles assumed they would always exist.

These are the canaries. If your role is on this list — or on the same page with a different job title that functions the same way — this chapter is for you.

— — —

Why the Middle Goes First

It is counter-intuitive. Most people expected the bottom of the org chart to go first. Entry-level roles always seemed the most vulnerable. Cheap, standardized, replaceable.

It is not happening that way, at least not yet. The middle is going first. The juniors are struggling to get hired at all, which is a different and equally serious problem. But the people actively being laid off in 2025 and 2026 are mostly in the middle.

Three reasons, in order.

The middle is where the AI actually competes. Entry-level work often involves physical presence, training context, institutional learning, and low absolute compensation. The cost of automating entry-level work is not much lower than the cost of a junior human, and it breaks the training pipeline a company needs for its own future. Middle-tier work, by contrast, is expensive, highly describable, and produces structured outputs AI is very good at. The ROI on automating the middle is enormous. The ROI on automating the entry level is marginal.

The top is protected by accountability. Chapter Five's Accountability Premium. Licensed, credentialed, signing professionals at the top of organizations cannot be replaced by AI, because AI cannot hold a license or accept personal liability. The more senior a role gets, the more the job tilts toward accountability functions that machines structurally cannot perform.

The middle is where the P&L optimization is visible. CFOs and VPs of Operations look at department costs and notice that the middle of the department is where the spend clusters. When they ask "can AI do this," the answer is most clearly yes for middle-tier work. The decisions get made mechanically, on a spreadsheet, often by people two layers removed from the actual humans being displaced.

None of these reasons are dramatic. There is no villain in this story. There are a lot of spreadsheets, a lot of quarterly earnings pressure, and a lot of mechanical logic that, applied consistently across thousands of companies, adds up to a generation-defining labor transition.

— — —

Dave's Team, Revisited

Remember Dave Palmieri from Chapter Two. The insurance underwriting manager outside Cleveland. Nineteen years at the firm. Fourteen people reporting to him when we met him.

Dave made his pivot. By 2026 he had rebranded himself as a Licensed Signer of Commercial Underwriting Decisions, leaned into his CPCU, and taken ownership of the firm's AI oversight process for commercial trucking risks in his region. His compensation went up twenty percent over two years. He is, by any normal definition, thriving.

His team did not all thrive.

Of the thirteen other people who were on his team in 2023, four are still there. Nine are not. Two took early retirement with decent packages and are fine. Three moved to smaller firms, took pay cuts of twenty to thirty percent, and are figuring out what is next. Two pivoted laterally into compliance and risk roles — not accidentally; they read the writing on the wall and did the work Chapter Twelve describes. One is doing contract underwriting work and hates it. One has been unemployed for fourteen months and has a kid in college and a mortgage that was calibrated to the salary he used to make.

Dave thinks about that last one a lot. They sat next to each other for seven years. They went to each other's kids' graduations. The guy was good at his job. The guy was not the problem. The job was the problem, and the job no longer exists.

This is the emotional weight of the cortex revolution at the ground level. It is not about a sinister AI. It is about specific humans with specific mortgages and specific kids, whose careers were predicated on economic assumptions that quietly stopped being true, who are now figuring out, mid-life, what the rest of their working life looks like.

This is real. The rest of this book does not help anyone if we do not acknowledge it is real.

— — —

Who Is Actually Making the Decision

One of the mistakes people make when they get laid off in this transition is personalizing the decision. They assume their manager turned on them. They assume the company did not value their work. They assume someone in HR had a specific grudge.

It is almost never any of those things. The decisions are mechanical and boring, and they come from a much less dramatic place than grievance.

The decisions originate in a CFO's office looking at quarterly P&L, noticing that a specific function's labor costs are not declining despite AI-driven productivity gains, and asking the function head why that is. The function head, not wanting to lose budget authority, does a headcount review. The headcount review identifies roles where AI productivity gains can support lower staffing. The decision cascades from there to HR, which implements the reduction, often without full context.

This is not personal. It is also not easy to fight at the individual level. You cannot out-work a spreadsheet. You can only change the spreadsheet's input by becoming the kind of worker whose output per dollar is high enough that you are a line item the CFO does not flag.

This is, again, the point of the first thirteen chapters. Physicality. Accountability. Taste. Adaptability. Combination. EQ. Orchestration. Solo-operator leverage. Those are not abstract virtues. They are the specific qualities that change how you look on the CFO's spreadsheet.

— — —

If You Are on the List Right Now

Let me speak directly to the reader whose role is on the autopsy list. You know who you are.

You probably have between twelve and twenty-four months before the decision reaches your specific role. That is not nothing. That is enough time to make a real pivot if you start this week. It is not enough time if you wait six more months to "see how things play out," because things are playing out right now, and the pivoters who started in 2024 are already ahead of you.

Here is the honest sequence. It is not glamorous. It works.

- **Accept the diagnosis.** Not abstractly. Specifically. Your specific role, your specific tasks, your specific trajectory. Write it down. Say it out loud to someone who can handle it. The denial phase is the most expensive phase of this transition, and the shortest path through it is to face it.
- **Choose a pivot target.** Use the framework from this book. Accountability-heavy Signer Role. Physicality-adjacent presence role. Taste-forward curation role. Agent-wrangler role. Compliance-native role. Pick one based on your actual skills, credentials, and temperament. Do not pick the most glamorous option; pick the most plausible one.
- **Identify the credential or skill gap.** What would you need — a certification, a portfolio, a specific project, a relationship — to be a plausible candidate for a pivot-target role in twelve months? Write that down. Make it specific.
- **Start closing the gap this week.** Do not wait for HR. Do not wait for your manager's permission. Do not wait for a clear signal. The clear signal is the package on your desk, and by then it is too late.
- **Keep your current job while you pivot.** Income continuity buys you options. Rage-quitting buys you six months of savings burn and a weaker negotiating position. Stay. Pivot. Leave when the new thing is ready.

This is not inspirational. It is practical. The people who successfully navigate these transitions almost always follow this sequence. The people who do not follow it almost always struggle.

— — —

Diagnostic

79. Is my current role on the list in this chapter? If not literally, is a role that does the same thing under a different title?

80. In the last six months, have I noticed any of the early signals — reduced workload, shortened one-on-ones, quiet hiring freezes on my team, my manager being asked to "think about efficiency"?

81. If I had to pivot in eighteen months, what would my target role be? Can I name it specifically?

82. What is the single biggest gap between where I am now and that target? Credential? Skill? Relationship? Portfolio?

83. What am I doing this month to close that gap?

———

Monday Morning

If you are on the list, and you have not started pivoting, start this week. Not next month. This week.

Open a document called "My Pivot." Write down your current role, honestly. Write down your pivot target, specifically. Write down the three biggest gaps between the two. Write down one action per gap that you will take this month.

Put a recurring thirty-minute block on your calendar called "Pivot work." Non-negotiable. Same time every week. Do the pivot work in that block.

Twelve months from now, that block of time will either be the reason you are okay, or the reason you wish you had started sooner. The math is brutal and straightforward.

— — —

Science Anchor

84. Goldman Sachs, *The Potentially Large Effects of Artificial Intelligence on Economic Growth* (March 2023). The 300 million FTE exposure estimate and the methodology behind it.
85. Anthropic Economic Index, updated periodically. Occupation-level real-world AI usage data.
86. McKinsey, *Generative AI and the Future of Work* series (2023–2025). The most detailed occupational analysis widely available.
87. BLS Occupational Employment and Wage Statistics, quarterly updates. The real-world data on where employment and wages are actually moving.
88. Daron Acemoglu and Pascual Restrepo, ongoing papers on automation, displacement, and the "reinstatement effect." The technical literature behind the trend data.

— — —

Close

A week after the severance announcement, Dave's former colleague calls him. It is a Saturday. The guy wants advice. He has been applying for forty jobs a month for three months. No offers. He is running out of savings. His wife is stressed. His kid just got into a second-choice college and tuition starts in the fall.

Dave listens for twenty minutes. He does not try to fix anything yet. He just listens, because that is the right move, and because the guy on the other end of the phone does not need a lecture about the EU AI Act.

When the guy is done talking, Dave asks one question. "What would you do for a living if you could start over at fifty-two?"

The guy does not have an answer. He has been applying to the same role, at every firm, for three months. He has not let himself ask that question. Nobody ever asks themselves that question unless they are forced to.

Dave tells him to take the weekend. Write the answer down. Call him Monday.

The guy calls Monday. He has an answer. It is not underwriting. It is something he has wanted to do for twenty years and has never given himself permission to consider.

Dave tells him: okay. Let's map it out. Here is how we start.

That is the end of this chapter, but it is the beginning of a different chapter in that guy's life. It starts the same week he got laid off. It starts with a question he was not asking until the work stopped asking it for him.

If this chapter is about you, take the weekend. Write the answer down. Call somebody Monday.

Next chapter: the other half of the labor transition. The blue-collar renaissance. Why Marisol is turning down work,

why the trades shortage is structural rather than cyclical, and why American culture got the value of physical work exactly backwards for half a century.

Chapter 15: *Blue-Collar Renaissance*

The plumber will save you. Not metaphorically. Literally.

It is a Saturday morning in Phoenix, one year after we last saw her. Marisol Quintero is at her kitchen table with a cup of coffee, her quarterly P&L open on her laptop, and her dog asleep at her feet. The numbers are good. The numbers are, honestly, a little embarrassing.

Revenue is up thirty-one percent year over year. Owner compensation is up eighteen percent. Her team has grown from twelve employees to sixteen, and she is actively hiring for three more positions she cannot fill. This week alone she turned down four commercial jobs, not because she did not want the work, but because she is booked six weeks out and her new customers can't wait that long.

Her biggest strategic question right now is whether to open a second location in Chandler, or whether to stay Phoenix-only and focus on higher-margin commercial work. She is working through it with her accountant. It is the kind of problem a lot of small-business owners would kill to have.

Marisol is not special. Marisol is the canary in the opposite coal mine. She is what the entire trades economy in America looks like in 2026: structural labor shortage, compressed time-to-profit, rising wages, expanding business ownership, and a pipeline of young workers that is nowhere close to what the economy needs.

This chapter is about why that is, and why it matters, and what you should do about it — whether you are a mid-career professional considering a pivot or a parent writing another tuition check.

— — —

The Shortage Is Structural

First, the numbers. The United States is short somewhere in the neighborhood of five hundred thousand skilled trades workers as of 2026, and the shortage is widening every year. The specific trades with the most severe gaps include plumbing, electrical, HVAC, welding, diesel mechanics, elevator mechanics, machinists, and increasingly, construction in general.

This is not a cyclical shortage. A cyclical shortage resolves itself in two to three years as wages rise, pull new workers in, and supply catches up to demand. A structural shortage is one where the underlying pipeline is broken, and rising wages cannot fix it, because the workers do not exist to be pulled in.

Three reinforcing factors make this a structural shortage.

Aging workforce. The median age of a skilled tradesperson in the United States is in the late forties. A substantial share of the workforce is within a decade of retirement. They are being replaced at roughly half the rate they are leaving. Every year that mismatch continues, the shortage widens.

Collapsed apprenticeship pipeline. The mid-twentieth-century apprenticeship system, which routed high-school graduates directly into the trades with structured multi-year training, was largely dismantled during the education-industrial-complex era of 1980 to 2010. Rebuilding it is happening, slowly, in some states and some trades, but not nearly fast enough to offset the retirements.

Cultural signal failure. For two generations, American culture aggressively signaled to ambitious young people that trade work was a consolation prize. Guidance counselors steered high performers toward four-year colleges. Parents steered their children the same way. The children who would have made outstanding electricians, plumbers, and welders were routed into business degrees and psychology majors instead. That miscasting is still being corrected, and it is being corrected twenty years late.

None of these is going to fix itself in the next decade. The trades economy is going to run hot — with all the attendant opportunities and stresses — through at least the 2030s, probably longer.

— — —

Real Wage Data (And Why You Should Believe It)

American culture's story about tradesman wages lags the reality by about fifteen years. Here is the current data.

The median licensed plumber in a major metro earns between ninety thousand and a hundred and thirty thousand dollars a year in salary-equivalent compensation, before any business ownership premium. The median master electrician earns in a similar range. HVAC journeymen, in markets where the demand is heaviest, earn comparable money. Welders working on specialty applications — pipeline, underwater, nuclear, aerospace — can clear two hundred thousand dollars a year in peak years.

Those are wages. The business ownership premium is separate and larger. A successful small-business owner in plumbing, electrical, or HVAC, running a crew of ten to twenty people, is often pulling three to five hundred thousand dollars a year in total owner compensation. This is not an outlier. This is a normal outcome for a competent tradesperson who builds a small business by their late thirties.

Compare this to the white-collar trajectory. A graduate with a mid-tier four-year degree, starting at fifty-five to seventy thousand dollars a year, reaches six figures around age thirty-five to forty in most fields. The trades worker, starting at twenty or twenty-one with minimal debt, reaches six figures in their mid-to-late twenties in most strong markets. The time-to-full-compensation is roughly a decade faster, and the debt load going in is close to zero.

Once you factor in the student debt drag on white-collar professionals — which delays home purchase, family formation, and net worth accumulation by roughly a decade — the trades worker's total lifetime economic trajectory is competitive with or better than most white-collar four-year-degree trajectories. And the gap is widening every year as the college-wage premium erodes under the cortex revolution.

This is not nostalgia for the days of hard hats and lunchpails. This is current, 2026 economic data. The trades are no longer a consolation prize. They are, in many markets, the premium path.

— — —

The Small-Business Piece

Here is the part that often gets missed. The most successful modern tradespeople are not, strictly speaking, tradespeople. They are small business owners whose business happens to be a trade.

My wife runs a professional handywoman business in Central Florida. She is not a factory worker with a hard hat and a lunch pail. She is a small-business owner who does direct client relationship management, quotes her own jobs, manages her own scheduling, handles her own marketing, owns her own equipment, sets her own rates, and chooses her own clients. Her business looks, structurally, much more like a solo consulting practice than it looks like the blue-collar caricature most of us inherited from mid-century sitcoms.

Most successful tradespeople in 2026 work this way. Either they own their own small business by their mid-thirties, or they work for a small business owned by someone in their peer group. The classic "W-2 employee of a giant construction firm" trajectory still exists, but it is not where the interesting money is. The interesting money is in ownership.

This means — and this is an important point — the tradespeople of the 2020s and 2030s need more than trade skills. They need business literacy. Customer relationship skills. Pricing and estimation judgment. Basic accounting and cash flow management. Marketing, increasingly digital. Hiring and crew management. Legal and insurance literacy.

This, not coincidentally, makes them excellent candidates to pair trade skills with AI leverage. A plumber who runs a small business and uses agent-driven systems for scheduling, quoting, customer communication, and compliance documentation is operating at a leverage level that sole-practitioner tradespeople never had before. The solo operator model from Chapter Thirteen applies directly. Marisol's competitors, five years from now, are going to be running their own agent fleets for the back-office work while they focus their human attention on the work that requires their physical presence and licensed judgment.

The Physicality Wall stacks with the Accountability Premium stacks with the Solo Operator Model. Three pillars of this book, all at once, in one profession. That is why the trades are arguably the single most defensible career path available to an ambitious American in 2026.

— — —

Why the Pipeline Is Broken (And Who Broke It)

I want to say this carefully, because it touches a nerve, and the nerve deserves to be touched gently.

For about forty years, the dominant narrative about American education assumed that a four-year college degree was the universal on-ramp to a stable middle-class life. High schools reorganized around this assumption. Parents oriented their ambitions around this assumption. Federal student loan programs subsidized this assumption. An entire industry of test prep, admissions consulting, and college counseling grew up around this assumption. By the early 2000s, anyone steering a kid toward the trades was treated, in most middle-class American contexts, as someone who had given up on that kid.

That narrative was wrong by about 2005. It is now wrong by a wide margin. And the infrastructure that made the narrative dominant is only slowly dismantling.

Which means there is a specific, structural mismatch in the American labor market: too many young people on the white-collar track facing a contracting job market, and too few young people on the trades track facing an expanding one. This mismatch will persist for another decade at least, because the cultural signals are slow to turn, and because the infrastructure of high school guidance, parental expectation, and social status has not caught up.

If you are a parent, you can fix a small piece of this inside your own family in one weekend. If you are a mid-career professional whose current path is uncertain, you can reposition yourself inside it in the next two years. Both moves are available. Both moves are being made, in large enough numbers to notice, but not in large enough numbers to close the gap.

— — —

Mid-Career Pivots into the Trades

Let me address the mid-career professional directly, because I know you are reading this with mixed feelings.

Yes, you can pivot into a trade at forty-five. I know people who did it at fifty. One who did it at fifty-eight. It is harder physically than it would have been at twenty. It is faster, in some ways, because you already know how to show up, work hard, communicate with clients, and run a small operation.

The on-ramp for most trades looks roughly like this. Research which specific trades have the most severe shortages in your geography. Identify the licensing pathway in your state — apprenticeship, journeyman, master. Find a reputable trade school or union apprenticeship program. Begin part-time if your current job allows it, or full-time if you have runway. Expect one to three years to journeyman competence, another two to three to master.

At the end of that period, in your mid-to-late forties or early fifties, you are a licensed tradesperson with many years of business and life experience that your twenty-year-old peers do not have. You are, structurally, one of the most hireable tradespeople in your market. The demand for your labor is not going to fall. The demand for your judgment is going to grow.

This is not for everyone. It is physically demanding. It is not glamorous in the pop-culture sense. It pays poorly during the training phase, and it requires investing money and time in retraining that most mid-career people are not able to invest. But for the reader who has the runway, the health, and the temperament, it is not crazy. It is, in fact, one of the cleanest pivots available to a mid-career professional whose current industry is contracting.

— — —

For Parents

If you have a kid in high school or in the first year of college, pay attention to this section. This is the one I want to make sure you read carefully.

Most parents, including parents who are nominally open to their kids considering the trades, have never actually taken their kid to visit a trade school or met with a working tradesperson about what the career looks like. The implicit assumption remains that college is the default and trades are the opt-out. This assumption is no longer serving most kids.

Here is the concrete ask. This month — this calendar month — take your teenager or young adult kid to visit one trade school in your area. Most of them do open houses on a regular schedule. Walk through the programs. Meet the instructors. Talk to second-year students about what their week looks like. Ask about placement rates, starting wages, licensure pathways, and career trajectories.

If the kid is already in college and miserable, ask the harder question. Is the degree they are pursuing actually going to deliver the career they want, given what they know about their industry's trajectory? Is there a trade that would use the strengths they actually have — spatial reasoning, mechanical aptitude, people skills — better than what they are currently studying? Is there a version of the plan in which they transfer into a trade program, or finish the degree and pursue a trade-adjacent business path?

You are not betraying your kid by asking these questions. You are parenting them through the largest labor transition of the twenty-first century. The parents who ask these questions openly, without judgment, are the ones whose kids end up okay. The parents who insist on the default path are often the ones whose kids are surprised at thirty.

I am not telling you to pull your kid out of school. I am telling you to help them see the full opportunity landscape, not just the one your guidance counselor grew up inside.

— — —

Diagnostic

89. If my kid or I were going to pivot into a trade in the next two years, what specific trade would it be and why?

90. What is the on-ramp in my state? Apprenticeship? Trade school? Union path? What is the timeline and cost?
91. Which trades in my geography have the most severe shortages right now, and what are the wage trajectories?
92. Do I know any working tradespeople well enough to have a candid conversation about what their career has actually been like? If not, could I make that a priority to get to know one this year?
93. If I were advising a nineteen-year-old today, what career path would I actually recommend? Is it different from the path I took?

— — —

Monday Morning

For parents: book the trade-school visit this week. Pick a Saturday this month. Go.

For mid-career pivoters: identify one trade with a structural shortage in your geography. Research the on-ramp. Visit one trade school or apprenticeship program in the next sixty days. Talk to one working tradesperson in that trade in the next thirty. No commitment required. Just information.

For everyone else: reconsider your mental model of what a successful American career looks like. The model most of us inherited — four-year degree, knowledge-work career, salaried employment — is still a path, but it is no longer the only respectable path, and it is no longer even the best-paying one for a substantial share of working adults. Update the model.

— — —

Science Anchor

94. BLS Occupational Outlook Handbook, skilled trades sections. Current wage data and ten-year projections, updated regularly.

95. Mike Rowe's mikeroweWORKS foundation and associated writing. A decade of advocacy that is finally, belatedly, being taken seriously.

96. Harvard Graduate School of Education, "Pathways to Prosperity" project. Long-running research on non-college career pathways in American labor markets.

97. Reports from the Home Builders Institute, Associated Builders and Contractors, and Associated General Contractors of America on the construction labor shortage. The industry literature is blunt about the gap and its causes.

— — —

Close

Marisol closes her laptop on Saturday afternoon. The P&L is filed. The decision on the Chandler expansion is not yet made, but she knows which way she is leaning. She takes her dog for a walk. Her spouse is inside, starting dinner. Her kids are in the pool.

In forty miles of every direction from where she sits, there are AI engineers worrying about their jobs, insurance underwriters worrying about their jobs, marketing directors worrying about their jobs. Marisol is not worried. Nobody is laying Marisol off. Nobody is going to lay Marisol off. Marisol is in a field that the cortex revolution, for all its power, cannot reach — because houses in

Phoenix will keep springing leaks and water heaters will keep failing and nobody has invented the robot that can crawl under the bungalow built in 1962 by a contractor who did not believe in building codes.

This is not a consolation prize. This is a career. This is, in many ways, the better career available to an ambitious American in 2026, and it has been hiding in plain sight while we kept steering our kids toward the career that is contracting.

If you have a teenager, take them to a trade school this month. If you are mid-career and unhappy, make the call to one tradesperson this week. If you are neither — if you are somewhere on the white-collar path and trying to figure out your next move — keep reading. We have five more chapters. Part Four continues by honoring the other careers that humans still own, and then we bring the whole book home in Part Five with the Survival Manual.

Marisol, for the record, will retire at her own pace. She may sell the business in her sixties. She may hand it to a daughter. She may keep running it. She will decide. That is the point of owning the kind of career the robots cannot take from you.

Next chapter: the first of three deep dives into the Care Economy. Healthcare, specifically. Why the doctor shortage is going to make the lawyer shortage look quaint, and why physicians — more than any other white-collar category — are about to have the decade of their lives.

Chapter 16: *Care Economy I - Healthcare*

The doctor shortage is going to make the lawyer shortage look quaint.

A few years ago I spent three years getting a specific headache disorder diagnosed. Cervicogenic migraine, if you are curious. Not a rare condition. Not a difficult diagnosis once the right specialist looks at it. The problem was not the disorder. The problem was that every specialist I needed to see had a six-to-nine-month waitlist, and by the time I saw one, the next referral was another nine-month wait, and the total diagnostic journey took longer than the problem should have required in a less backlogged system.

I am one of tens of millions of Americans currently waiting for medical care we will eventually get, more or less, but slowly. Some people reading this sentence are waiting for something less patient than a headache diagnosis. Some are waiting for cancer screening. Some are waiting for mental health care. Some are waiting for a primary care doctor to take them on as a new patient and being told the answer is no, not accepting new patients.

This is not an inefficiency to be optimized. This is a labor shortage at the scale of a public crisis. It is also, for anyone willing to enter the field, one of the largest and most durable economic opportunities of the next twenty years.

Welcome to Part Four's three-chapter deep dive into the Care Economy. We start with healthcare, because healthcare is biggest.

— — —

The Numbers (Which Are Grim Unless You Are Hiring)

The Association of American Medical Colleges has been projecting physician shortages for over a decade. Their current projections put the 2034 U.S. physician shortage somewhere between 37,000 and 124,000 doctors, depending on specialty mix and demand assumptions. Those are the official numbers. Most working clinicians will tell you the real number is higher, because the AAMC's models assume patients who cannot access care simply go without, which understates real demand.

The nursing shortage is worse. The U.S. is short approximately 200,000 registered nurses per year, with the gap projected to widen through the 2030s. Schools of nursing turn away qualified applicants annually, not because they do not want to train them, but because there are not enough nursing faculty to teach them. The pipeline is choking on its own throughput.

Allied health fields — radiologic technology, ultrasound, MRI, physical therapy, occupational therapy, respiratory therapy, pharmacy technicians, medical lab technology — are short workers at rates that make the physician shortage look manageable by comparison. Hospitals routinely cannot staff specific departments. Outpatient imaging centers have six-to-twelve-week appointment backlogs in most metros. Retail pharmacies have been reducing hours because they cannot hire enough pharmacists.

Mental health care has its own staffing crisis, severe enough to warrant its own chapter, which is next. Geriatric care has its own crisis too, which is the chapter after that.

All of this is happening while demand for healthcare services is rising — because the population is aging, because chronic disease burden is up, because mental health awareness has broadened the diagnostic footprint, and because modern medicine keeps people alive longer with more conditions to manage.

Supply is flat to falling. Demand is rising. The gap is the opportunity.

— — —

Why AI Makes Doctors More Valuable, Not Less

You might have expected, back in 2022, that AI would ease the doctor shortage. Scan reading would be automated. Diagnosis would be assisted. Chart review would be accelerated. Surely this would reduce the need for physicians.

It has done the opposite. AI in medicine, over the last three years, has made existing physicians dramatically more productive — able to see more patients, make more decisions, handle more throughput in the same hours. In a labor market with fixed supply and rising demand, productivity gains do not reduce headcount needs. They just allow existing physicians to meet more of the gap, while the gap continues to widen faster than productivity can close it.

Meanwhile, nothing about AI has displaced the physician as the legally and clinically accountable decision-maker. The reasoning from Chapter Five applies in its most concentrated form here. The physician is the signer. The physician holds the license. The physician has the malpractice exposure. The physician takes the phone call when something goes wrong at 3 a.m. None of those functions transfer to software, and none of them will for the foreseeable future, because the regulatory stack and the liability framework are specifically designed to keep them in human hands.

The same principle applies to nurse practitioners, physician assistants, and licensed pharmacists. Every allied health profession with a license and a scope of practice operates under the same structural protection. AI expands their capacity. Regulation requires their presence. The combination creates a rising-value career in a field that was already resilient.

If you pivoted into a medical career in 2020 thinking it was a safe choice, you were right. You were more right than you knew. Your career trajectory over the next two decades is going to be better than almost any other white-collar path available in American labor markets, and the delta is growing.

— — —

The Allied Health Wave

Most of this chapter's readers will not become physicians. The pathway is too long, the debt is too punishing, and the training pipeline is too narrow. That is fine. Physicians are one percent of the healthcare workforce. The other ninety-nine percent is where most of the actual hiring happens, and a lot of it is accessible to mid-career pivoters.

The specific allied health fields worth understanding in 2026, by training time and typical salary trajectory:

- **Registered nurse (RN).** Two-to-four-year training, depending on pathway. Median pay nationally in the seventy-to-ninety-thousand-dollar range, with higher compensation in specialty units. Demand is severe and widening. Nursing shortages are projected through the 2030s at minimum.

- **Nurse practitioner (NP).** Requires RN plus master's degree, typically six-to-seven years total. Practices with significant autonomy. Compensation in the six-figure range, often much higher in underserved specialties. Specialty pathways include family practice, psychiatric-mental health, acute care, geriatrics.
- **Physician assistant (PA).** Two-to-three-year master's program after prerequisites. Works closely with physicians across most specialties. Compensation comparable to nurse practitioners, in the six-figure range. Shortage is severe.
- **Pharmacist.** Six-year PharmD program. Compensation in six figures. Demand is strong in specialty and hospital pharmacy, more mixed in retail due to chain-store consolidation pressures, but overall the profession is short workers.
- **Radiologic technologist, MRI technologist, sonographer.** Two-to-three-year programs, often at community colleges. Starting compensation in the high five figures, rising to six figures with experience and specialty certifications. Openings outstripping applicants in most metros. This is one of the most accessible pivots in the entire healthcare sector.
- **Physical therapist.** Requires DPT, typically seven years total. Compensation in the upper five figures to low six figures. Demand strong, particularly in aging-related rehabilitation.

- **Occupational therapist.** Master's or doctorate required. Compensation and demand profile similar to PT.
- **Respiratory therapist.** Two-year associate degree typically sufficient for entry. Compensation in the high five figures to low six figures. Heavily hired, especially in hospital settings.
- **Medical lab technologist.** Two-to-four-year programs. Lower visibility, steady demand, meaningful work.
- **Speech-language pathologist.** Master's required. Strong demand, particularly in schools and rehabilitation settings.

Someone close to me spent the last year researching several of these specific pathways — radiologic technology, MRI, and ultrasound, specifically — as a possible career pivot. The numbers, when she ran them carefully, looked better than any white-collar pivot she could have considered. Two years of training at a community college. Starting wages in the sixty-to-eighty-thousand-dollar range depending on local market. Ten-year wage projections that outpace almost every comparable white-collar track. Job security rooted in the Physicality Wall, the Accountability Premium, and the underlying shortage.

This pattern holds across the allied health sector. If the white-collar career is cracking and the full physician pathway looks too long, the allied health on-ramp is shorter, cheaper, and opens into a field where nobody is being laid off.

— — —

The Two-Track Choice

A useful frame for thinking about healthcare pivots is the two-track choice.

Track one: the clinical track. You actually touch patients. You deliver care. Whatever role you train into, you are in the room with humans, doing work that requires your body, your hands, and your licensed judgment. Nursing. NP. PA. Imaging tech. PT. OT. Respiratory. These roles sit squarely at the intersection of the three pillars — Physicality, Accountability, and Presence.

Track two: the non-clinical healthcare track. You support clinical work without delivering it. Healthcare administration. Revenue cycle management. Medical coding (until AI eats it). Health informatics. Clinical research coordination. Compliance within healthcare. Care navigation. These roles are in the industry but do not require direct patient care. They are better for readers with clinical interest but limited physical or scheduling flexibility.

Both tracks are growing. The clinical track pays better per year of training and is more directly protected from automation, because the Physicality Wall is load-bearing in clinical work. The non-clinical track is more accessible to mid-career professionals with transferable skills and fewer physical requirements.

If you are serious about pivoting into healthcare, you need to pick a track before you start the specific program, because the pathways diverge quickly and backing up mid-training is expensive.

— — —

The Accountability-Physicality Stack in Healthcare

Earlier chapters introduced the three pillars — Physicality, Accountability, Taste — as the things that stay human. Healthcare is the single best example of what happens when those pillars stack on top of each other in one career.

A nurse in a hospital does physical work with patients. That is the Physicality Wall. The nurse holds a license, signs assessments, and can be personally sanctioned for failures of care. That is the Accountability Premium. The nurse reads moods, notices signals, provides comfort, and knows when to escalate. That is the EQ Economy from Chapter Nine. The nurse works inside a system of agent-assisted charting, decision support, and patient monitoring that the nurse supervises rather than executes. That is the new literacy from Chapter Ten.

All four pillars, in one job. That is why nursing compensation has been rising in real terms throughout the cortex revolution, and why it will keep rising. The market is pricing what it cannot substitute.

The same stack applies, with different emphases, across most clinical roles. Physician: physicality, accountability, taste. Physical therapist: physicality, accountability, presence. Imaging tech: physicality, accountability, system supervision. Each role sits at some specific intersection of the pillars, and each intersection is exactly where the human premium concentrates.

If you built a pure AI-resistance scorecard for American careers in 2026, clinical healthcare would sweep the top ten slots.

— — —

Diagnostic

- Have I seriously considered a clinical healthcare career, or have I assumed it is not for me without actually doing the math?

- What is the shortest allied health pathway that fits my existing skills and physical capacity? Have I looked it up for my state?
- If I were to pivot, could I do it without leaving my current job — starting with a community college program part-time while keeping income continuity?
- Am I more drawn to direct patient care or to supporting roles around the clinical work? Both are valid; the answer shapes the pathway.
- If I have a child considering post-secondary education, have I walked them through the allied health option explicitly, or defaulted to the four-year-degree pathway?

— — —

Monday Morning

Pick one allied health pathway this week and do the actual research. Visit the website of one community college or university in your area. Look at the specific program's prerequisites, duration, cost, and placement outcomes. Call the admissions office with three questions.

If the numbers look reasonable, visit an information session in the next sixty days. No commitment. Just information.

If you already know the pathway you want, enroll. Not next semester. This one. The opportunity cost of another year of uncertainty is bigger than the cost of starting.

— — —

Science Anchor

- AAMC physician workforce projections, published biennially. The authoritative source for U.S. physician shortage data.
- BLS Occupational Outlook Handbook, healthcare and technical occupations sections. Detailed projections by role through 2034.
- HRSA workforce projections for specific allied health roles. Worth reading for the state-level breakdowns.
- American Association of Colleges of Nursing annual reports. The clearest data on the nursing workforce and pipeline issues.
- State licensing board pages for the specific pathway you are considering. These are the official source of truth on requirements and timelines.

— — —

Close

I walked out of my specialist's office a couple of years ago with a diagnosis, a treatment plan, and a standing weekly appointment. The care, once I reached it, was excellent. The specialist was kind, thorough, and spent more time with me than any doctor I had seen in a decade. She was also, she told me offhand, booked through the next six months, and her practice was capping new patients because they physically could not keep up.

Every one of those patients waiting for her is somebody's mother, somebody's husband, somebody's kid. Every one of them needs care that is, at the bottleneck, staffing-limited.

If you are considering a career pivot in 2026, and you are physically able, and you have the patience to train, healthcare is the single most reliable bet available to you. The demand is structural and rising. The jobs are protected from automation by law, physics, and liability. The work is meaningful in a way most careers cannot match. The compensation is holding and in many sub-fields rising in real terms.

Nobody is laying anyone off in American healthcare. Everyone is hiring. Everyone will keep hiring, for the next twenty years at least.

You could be one of the people they hire.

Next chapter: why the mental health workforce crisis is even more severe than the physical health one, why Eleanor is still booked eighteen months out, and why becoming a therapist, a counselor, or a credentialed coach is one of the most durable career decisions a mid-career American can make in 2026.

Chapter 17: *Care Economy II: Mental Health and Coaching*

It is a Tuesday afternoon in Boston. Dr. Eleanor Voss has a new patient referral on her desk. A twenty-six-year-old man. Severe panic disorder. His primary care doctor flagged the case as urgent.

Eleanor cannot take him. Her practice has been closed to new patients for two years. She calls three colleagues she trusts. One is not accepting new patients. One is booked into next February. The third could see him in about seven months.

Eleanor calls the primary care doctor back. Seven months is what she has. The doctor sighs. She will try to find something sooner. Both of them know there is no "sooner." The entire system is gridlocked, from community clinics up through private practice, and the situation is worse every quarter.

This is not a failure of the mental health system. The mental health system is working as designed, within the constraints it has, applying the practitioners it has. The failure is that the practitioners do not exist in anything close to the numbers America needs. Every working clinician has a waitlist. Every waitlist has waitlists behind it. The people behind all of those waitlists are your neighbors, your coworkers, your children.

If you have been listening to the argument of this book, you already see the implication. A structural, widening labor shortage in a field the cortex revolution cannot reach is a secular tailwind. A secular tailwind in a meaningful profession is the closest thing to a guaranteed career available in 2026.

Let us talk about mental health work.

— — —

The Scale of the Gap

The United States is short, by most serious estimates, at least a hundred thousand mental health professionals. Some estimates are considerably higher, depending on how you count providers and what you include as unmet demand. Roughly one in three American adults experiences a diagnosable mental health condition in a given year. The share of those who actually receive treatment hovers around half, and in lower-income and rural areas it is dramatically worse.

The demand curve accelerated during the 2020 pandemic and has not come back down. Young adult demand specifically has climbed steadily since 2018, tracking the data in the Murthy loneliness advisory discussed in Chapter Nine. The share of college students seeking mental health services has roughly doubled over the last decade. Pediatric and adolescent mental health demand has outstripped every projection made by workforce planners in the last twenty years.

On the supply side, the workforce is aging. A substantial share of working psychologists and clinical social workers are within a decade of retirement. Training programs have expanded, but not nearly fast enough to close the gap. The psychiatrist population in particular has been growing at under one percent a year while demand has grown at double digits.

The gap compounds. The patients who cannot access care in their twenties get sicker. The patients who get sicker take more treatment time per session. The treatment time per session eats into capacity. The capacity falls behind demand even faster. Every year the system gets further behind itself, and nobody expects this to self-correct inside the decade.

This is, coldly stated, the largest and fastest-growing professional labor shortage in American knowledge work that does not require a physician's license.

— — —

Why AI Does Not Solve This

We covered the uncanny intimacy valley in Chapter Nine. AI companions and AI therapy apps can produce short-term engagement and occasional genuine benefit, but they hit a durable ceiling on retention and outcomes. Users notice, eventually, that they are being met by a mirror rather than a person. The therapeutic effect requires a witness who is actually present, and a witness that could choose not to be.

There is an additional layer specific to mental health care that is worth naming. The most effective therapeutic work happens across a multi-year relationship, in which the clinician comes to know the patient's history, patterns, triggers, relationships, and narrative in depth. AI systems, as currently architected, cannot sustain that kind of relationship at the level of a trained human clinician, for reasons both technical (context limits, memory constraints) and clinical (the therapeutic alliance is a specific human relational phenomenon, not an information exchange).

AI is going to keep expanding in adjunct mental health roles — intake triage, skills training between sessions, symptom tracking, basic CBT exercises, crisis screening. These are real and useful. They do not replace the licensed clinician. They give the clinician leverage, exactly the way AI gives other knowledge workers leverage, expanding their practice capacity without substituting for their role.

Which means — as with physicians — the productivity gains from AI in mental health care flow directly into expanded demand matching, not into displacement of practitioners. The clinician with an agent-assisted practice sees more patients, not fewer, at compensation that keeps pace with or exceeds the productivity gain.

— — —

The Pathways

Mental health work has multiple on-ramps at multiple training durations. This is unusual for a high-value field and it is one of the reasons this chapter warrants close attention from mid-career readers specifically.

- **Licensed clinical social worker (LCSW or LICSW).** Master's degree in social work, typically two years full-time, plus post-graduate supervised clinical hours to license, typically two more years. Total time to license: about four years. Scope of practice includes individual, family, and group therapy. Compensation ranges widely by setting; established private practice in a major metro can clear six figures comfortably, with full schedules. This is the single most common clinical mental health licensure in the United States.

- **Licensed marriage and family therapist (LMFT).** Master's degree in marriage and family therapy, plus supervised clinical hours. Similar total time-to-license as LCSW. Specializes in relational and family work. Strong demand, particularly in couples and family contexts.
- **Licensed professional counselor (LPC), sometimes called LMHC.** Master's degree in counseling, plus supervised hours. Similar pathway. Broadest scope of the three master's-level licenses. Licensing specifics and scope vary by state.
- **Clinical psychologist (PhD or PsyD).** Doctoral program, typically five to seven years, plus supervised hours. Longest pathway, highest earning ceiling among non-prescribing clinicians, strongest research and assessment scope.
- **Psychiatric nurse practitioner (PMHNP).** RN plus psychiatric-mental-health nurse practitioner master's. Can prescribe medications in most states. This is one of the most important and growing credentials in the entire system, because the severe psychiatrist shortage has shifted much of the medication-management work to PMHNPs.
- **Psychiatrist.** MD plus residency, eight-plus years of training. Highest cost of entry, highest earning ceiling, most severe shortage. For early-career readers with the runway and the inclination, this is an extraordinary bet. For mid-career pivoters, the math usually does not work.

The master's-level clinical licenses — LCSW, LMFT, LPC — are the sweet spot for most mid-career pivoters. Four years. Accessible programs, many of which are part-time or hybrid. Immediate post-graduation hireability, even during supervised-hours stage. Full licensure opens into private practice, group practice, community clinics, and increasingly, tele-therapy platforms.

If you can tolerate the training time and the specific emotional labor of clinical work, this is a career that will not stop needing you.

— — —

The Coaching Economy

Adjacent to clinical mental health is the coaching economy, and it deserves a careful treatment because the coaching field is both enormous and variable in quality.

Coaching, at its best, is a structured helping relationship in which a trained coach supports a client in articulating goals, identifying obstacles, and working through action plans. Evidence-based approaches exist. Formal credentialing exists. Research supporting the effectiveness of coaching in specific applications — executive development, career transition, performance improvement, life-stage transitions — is real.

Coaching at its worst is Instagram pseudo-therapy from people with no training and no credentials charging three hundred dollars an hour to clients they should be referring out. The field is not regulated in most jurisdictions, which means the good, the mediocre, and the actively harmful all compete in the same market.

If you are considering coaching as a career, take the field seriously. The credentialing pathways that matter:

- **International Coaching Federation (ICF).** The dominant credentialing body. Three levels — ACC, PCC, MCC — with increasing training and experience requirements. ICF credentials are recognized and respected in corporate coaching markets and by sophisticated individual clients.
- **Board Certified Coach (BCC).** Offered through the Center for Credentialing & Education. Widely accepted in certain settings.
- **Specialized credentials.** NBHWC for health and wellness coaching. Various therapy-adjacent credentials in specific modalities.

Credentialed coaches with established practices in major metros typically charge two hundred to five hundred dollars an hour, often more. A fully booked coaching practice of twenty to twenty-five client hours a week produces a substantial income, with structure flexibility that most clinical careers cannot match.

The downsides are real. Coaching does not address serious mental health issues, and a good coach has to recognize when a client needs a clinician instead. Coaching is not insurance-reimbursable, which limits the client base. Coaching is uncredentialed in most states, which means the clients you most want to attract may be skeptical of the field generally and may require you to demonstrate credibility more aggressively than licensed clinicians have to.

If you are drawn to helping work but not to the clinical training pathway, coaching — done with real credentials, real training, and real integrity — is a legitimate career. If you are thinking about it as a shortcut because the clinical training looks too hard, please do not. The field already has too many people who took that shortcut.

— — —

The Temperament Question

Mental health work is not for everyone. I want to say this plainly, because every chapter in this book ends with an action step, and a reader could charge into a four-year LCSW program without asking whether they actually want to do the work.

Do you want to do the work? Specifically:

- Can you sit with another person's suffering without trying to fix it?
- Can you hold confidentiality around things that would keep other people awake at night?
- Can you tolerate clients who do not improve, or who improve and then regress, or who end therapy abruptly?
- Can you take on a load of twenty-five-plus emotionally heavy sessions a week without carrying them home?
- Can you do this work for decades, not for a year?

The people who thrive in mental health careers have usually answered yes to all of those, not by virtue of natural temperament but by virtue of training, supervision, and a lot of their own therapy. If you are unsure whether this is you, consider an entry-level role — a community mental health paraprofessional position, a peer support role, a crisis-line volunteer shift — before you commit to the graduate program. A month of close contact with the actual work tells you more than a year of reading about it.

— — —

Diagnostic

- Do people already come to me with their problems, unbidden? Or do they avoid emotional topics with me?
- When I think about doing mental health work as a career, do I feel drawn or exhausted? Be honest.
- If I were going to train in this field, which specific pathway fits my temperament, my finances, and my time horizon?
- Do I know anyone working in mental health whom I could shadow or interview for a few hours in the next month?
- Am I romanticizing the field because I am burned out on my current one, or am I genuinely suited to this kind of work?

— — —

Monday Morning

If this field draws you, spend this month shadowing. Volunteer for a crisis line. Attend an open information session at a local master's in social work program. Interview two working clinicians about their actual week. The real question is not whether the field is a good career — it is. The real question is whether you are suited to do the specific work.

If the month confirms the draw, begin the application process for one specific program. Not five. One. The specificity forces you to decide. If the month reveals that the draw was more about escape than suitability, you have learned something valuable without paying tuition for it.

— — —

Science Anchor

- HRSA's workforce projections for mental health professionals. Detailed gap data by state and specialty.
- SAMHSA annual National Survey on Drug Use and Health. Demand-side data for mental health services.
- APA workforce reports and National Association of Social Workers member surveys. Practitioner-side data on earnings, demographics, retention.
- The Murthy advisory on loneliness (2023). Still the clearest articulation of the demand environment that is feeding the mental health workforce crisis.
- International Coaching Federation research on coaching efficacy and market size. Relevant if the coaching pathway appeals to you.

— — —

Close

Eleanor closes out her day at 7:15 p.m. on Tuesday. She has one voicemail left to answer — a former patient who wants to refer her own daughter for a consult. Eleanor listens twice. She writes down the name. She will call in the morning and explain, gently, that she cannot take the daughter either, and she will do what she does with every referral she cannot personally take: she will spend twenty minutes on the phone trying to find someone who can.

That twenty minutes, repeated across the working lives of tens of thousands of overworked clinicians, is what the American mental health system currently runs on. The clinicians cannot stop. The patients keep coming. The only thing that fixes this, structurally, is more clinicians.

Which is to say: if you do this work, you will never, in your working life, run out of people who need you. You will never have to worry about being displaced. You will never, when your employer announces a restructuring, be on the list. You will, however, have to do the work — which is not small, not easy, and not glamorous.

If that is you, we need you. The system is waiting.

Next chapter: the demographic wave nobody wants to talk about. Ten thousand Americans turning sixty-five every day, for a decade running. Somebody has to care for them. That somebody is, increasingly, not going to be their adult children. Welcome to the single largest labor shortage in American economic history.

Chapter 18: *Care Economy III: The Silver Tsunami*

By 2030, more Americans will be over sixty-five than under eighteen. This has never happened before in U.S. history. Not once. Not ever.

I want you to sit with that sentence, because if you did not sit with it, you would treat it like every other demographic factoid that gets cited on a Wednesday and forgotten by Friday. This is not a factoid. This is the single most important structural force shaping the American labor market of the 2030s. It is bigger than AI. It is bigger than the trades shortage. It is bigger than the mental health crisis. It is, in fact, the thing that is going to interact with all of those other forces and produce the specific labor market you and your children are going to be working inside.

Ten thousand Americans turn sixty-five every single day. This has been happening since 2011. It will keep happening until roughly 2030. By the end of that window, roughly seventy-one million Americans — more than one in five — will be senior citizens.

Many of them are going to need care. Some will need a little. Some will need a lot. Cumulatively, they are going to require a workforce that does not currently exist.

This chapter is about who is going to provide that care and how you might become one of the people providing it — either for wages, or as a business owner, or to support your own family members, or all three at once.

— — —

The Math Is Uncomfortable

Here is the brief version of the demographic math.

The ratio of potential caregivers to older Americans — the Caregiver Support Ratio — has been declining for decades. In 2010 there were roughly seven potential family caregivers, aged forty-five to sixty-four, for every American eighty or older. By 2030, that ratio is projected to be about four to one. By 2050, closer to three to one. The ratio is falling because people are having fewer children later, spreading siblings thinner, and living longer.

Paid caregiving fills the gap that family caregiving cannot. Home health aides, certified nursing assistants, personal care aides, hospice workers, geriatric care managers, geriatric specialist physicians and nurse practitioners. Every one of these roles is facing a shortage. The Bureau of Labor Statistics projects home health and personal care aide employment alone will need to add roughly 800,000 new workers in the next decade, making it one of the largest single-occupation gaps in the entire U.S. economy.

Meanwhile, the medical subspecialties that care for older adults — geriatric medicine, palliative care, hospice — are among the least-chosen specialties by new physicians, because they have historically paid less than procedural specialties. That pay gap is now inverting. Geriatric-trained physicians in 2026 are being recruited with signing bonuses and compensation packages that did not exist five years ago, precisely because demand has outstripped the small training pipeline.

The math is uncomfortable because the numbers are too big to solve with any single intervention. Immigration can help, and has helped; home health is disproportionately staffed by immigrants, and restricting immigration disproportionately worsens the caregiver shortage. Training programs can help. Wage increases can help. None of them, alone, closes the gap. The gap will be managed, badly, through the 2030s, and the people who can do the work will be paid more every year as the shortage deepens.

———

What the Silver Tsunami Actually Looks Like

Most readers of this book either have aging parents already or will in the next decade. If you have not yet been through the experience of finding care for an aging parent, let me describe what it is like in 2026, because the reality does not match the image.

A typical scenario. Your mother, seventy-nine, lives alone. She has been managing fine. Then she has a fall. A hip fracture. Surgery. Rehab. She is discharged to her home with instructions for six weeks of physical therapy and in-home assistance.

You call the first home care agency. Their next available in-home aide is two weeks out, and only for fifteen hours a week, not the forty the discharge plan recommends. You call the second agency. Same answer. You call the third. Same. You end up taking time off work to be with her during the gap, while managing your own job and your own kids. This is the sandwich generation experience, and it is getting tighter every year.

A slightly different scenario. Your father, eighty-two, has moderate dementia. He can still live at home with significant help, but the help has to be consistent. You start calling memory care facilities. The average waiting list is four-to-eight months. The average monthly cost, at the facilities you would actually let your father live in, is seven to ten thousand dollars. Medicare does not cover it. Medicaid covers it only after you have spent down the assets. You begin doing spreadsheets at 2 a.m. trying to figure out how this works financially for a family that considered itself reasonably prepared for retirement.

Both scenarios are extremely common. Both are about to become more common, because the demographic wave is cresting and the supply of care is flat. Every American family will face some version of this in the next decade, and the experience of navigating it is going to convert millions of Americans — formerly outside the care economy — into people who now think about it constantly.

Which includes you, probably, if you have not already gone through it.

— — —

The Career Landscape

Roles in the eldercare economy range from short-training entry-level positions to long-training specialist physician roles. The range matters because it makes this sector accessible as a career pivot at almost any life stage.

- **Home health aide / personal care aide.** Training requirements vary by state and employer; minimums are often measured in weeks, not years. Median wages have been rising steadily as the shortage deepens. Work is physically and emotionally demanding. Turnover is high. Readers considering this role should understand both the accessibility and the demands.

- **Certified nursing assistant (CNA).** State-certified training programs, typically four-to-twelve weeks. CNA is the backbone role in nursing homes, assisted living, and increasingly in home care. Wages higher than home health aide, scope of practice broader, career ladder opens into LPN, RN, and beyond.
- **Hospice aide / hospice worker.** Specialized training beyond CNA for end-of-life care. Demanding emotional work; also some of the most meaningful work in the entire healthcare sector. Chronic shortage.
- **Geriatric care manager.** Often a licensed social worker or nurse with additional specialization. Coordinates care for older adults and families navigating the system. Private practice possible. Demand growing faster than training capacity.
- **Geriatrician (physician).** Internal medicine or family medicine residency plus geriatric fellowship. One of the most acutely short-staffed medical specialties in the country. Signing bonuses, loan forgiveness, and competitive compensation packages increasingly common.
- **Geriatric nurse practitioner (GNP).** RN plus master's degree with geriatric specialization. Can provide substantial clinical care autonomously in most states. Growing fast. Worth serious attention for RNs considering advanced practice.
- **Eldercare business ownership.** Home care agencies, adult day programs, senior move managers, specialized home-modification businesses, senior-focused fitness programs, memory-care concierge services. This is where a

lot of the most interesting entrepreneurial action is going to happen in the next decade.

Notice again the stacking of the three pillars. Most of these roles involve physical presence, licensed accountability, and direct emotional presence with another human. Every eldercare role is resistant to automation by construction, and the demand side is guaranteed for at least the next twenty years by demographic forces that are already in motion.

— — —

The Sandwich Generation Career

One specific pattern is worth naming, because it fits a substantial share of this book's readers.

If you are between forty-five and sixty-five, you are probably already a caregiver for someone aging in your family, even if you do not call yourself that yet. Helping a parent with Medicare paperwork. Driving them to appointments. Researching insurance options. Managing medications. Negotiating with home care agencies.

That work is, in fact, care management. The skills you are already developing — navigating insurance, coordinating between medical providers, advocating for elderly relatives who cannot advocate for themselves, managing the emotional load of caregiving while holding down a job — are the exact skills a professional geriatric care manager uses, credentialed.

Which means, for a meaningful share of mid-career readers, the pivot into the care economy is not a pivot at all. It is a credentialing of work you are already doing. A certificate program in aging services, or a credential as a geriatric care manager, or a state-specific care navigator credential, converts unpaid family caregiving experience into paid professional services. The clients are not hard to find; they are every family in your community facing what you have been facing.

I have known multiple people in their fifties who made this pivot. The ones who did it intentionally — credential, marketing, setting up the business as a professional practice — are doing well. The ones who drifted into it without formalizing are often burning out without income. The difference is the deliberate step of turning the experience into a profession.

———

What This Means for Your Retirement

A brief detour that is relevant to this chapter and to the next one.

A lot of this book's readers are asking, quietly, what retirement is going to look like for them. The short answer is: different from what it looked like for your parents. The demographic wave that is creating the silver tsunami is also shifting how retirement gets financed, staffed, and lived. Social Security and Medicare are going to hold, but the margins around them are going to be tighter. The quality of the care you can access is going to depend on what you can personally afford and who you know in your community. The gap between well-resourced and under-resourced aging is widening fast.

For your own retirement planning — and I am not a financial advisor, so please take this as conceptual framing rather than specific advice — the implication is that you probably need a larger cushion for healthcare and long-term care expenses than conventional retirement planning assumed ten years ago. We will go deeper into this in Chapter Twenty. For now, note that the silver tsunami is not only a labor market reality. It is a personal planning reality, and the two interact.

— — —

Diagnostic

- Do I have an aging parent or relative whose care needs I am currently navigating? How many hours a week does that take?
- Have I researched what care costs in my community — home care rates, assisted living rates, memory care rates? Do I know what Medicare covers and what it does not?
- If I were going to pivot into an eldercare profession, which specific role best fits my training tolerance, physical capacity, and temperament?
- Is there a professional credential I could earn, in the next two years, that would formalize caregiving work I am already doing for my own family?
- If my parents needed care in the next year, do I have a plan, or would I be improvising in a crisis?

— — —

Monday Morning

If you already have aging parents, this week, have the conversation with them about what they want as they age. Not the abstract conversation. The specific one. Do they want to age in place? What would that require? Who would do it? Who pays?

If you do not have aging parents but you can see this career opportunity, pick one specific role from the list above and research the pathway in your state this week. CNA programs in your area. Geriatric care manager certification through the Aging Life Care Association. Home care agency franchise economics.

The care economy is the biggest labor opportunity in America over the next twenty years. Most people reading this book will end up inside it somehow — as workers, as business owners, as family caregivers. The deliberate path is cheaper than the drift path.

— — —

Science Anchor

- U.S. Census Bureau population projections, aging sections. The primary source on the demographic math.
- BLS Occupational Outlook Handbook, home health and personal care aide, CNA, and related entries. Employment projections through 2034.
- AARP and the National Alliance for Caregiving joint surveys on family caregiving. The data on the sandwich generation experience.
- Alzheimer's Association annual Facts and Figures reports. Detailed data on the specific dementia caregiving burden, which is one of the fastest-growing segments of the overall caregiving challenge.
- Aging Life Care Association materials if the geriatric care manager pathway interests you. Credentialing body and professional standards.

— — —

Close

My father was a solid guy. Not a complicated man. He worked a steady job for decades. He did not plan for aging care in any structured way, because his generation mostly did not. When he needed help at the end, my mother did most of it, and then my siblings and I picked up whatever she could not.

That pattern — the family absorbing the care burden through unpaid labor — held for about four generations of Americans. It will not hold for the generation currently aging, because the demographic math does not allow it. There are not enough adult children per senior. The adult children who exist are working two jobs and raising kids further apart in age than their own parents did. The burden cannot be absorbed at family scale anymore.

Which means it will be absorbed professionally, at wages that are going to keep rising until the workforce catches up with demand, which will not happen in our lifetimes.

If you can do this work, we need you. If you can run a business that provides this work, we need that too. If you can neither, you will still, probably, pay for it at some point in the next decade — for your own parents, for your spouse, or for yourself.

The care economy is the single largest unseen labor opportunity in the American economy in 2026. Most people have not noticed yet. The ones who notice early have a decade of head start before the rest of the market catches up.

Next chapter: we lean forward into the speculative but evidence-based. Jobs that do not exist yet, but will within five years. How to position yourself for roles that have not been invented. How to be the person hired for the position

the company did not know it needed until you showed them what was possible.

Chapter 19: Jobs That Do Not Exist Yet

In 1995, the job title "social media manager" did not exist. Nobody had it on their business card. Nobody had ever been hired into the role. The concept did not make sense because the platforms that the role would eventually manage had not been invented.

By 2008, social media manager was a real job title. By 2015, it was a standard role at every mid-to-large company in the United States. By 2022, there were about a quarter million people in America with that specific job title, and many more whose actual work was social media management under a different title.

Zero to quarter-million, in under twenty years, for a job nobody had heard of in 1995.

This pattern is not unusual. It is, in fact, how most of the best-paying careers of any given decade come into existence. UX researcher. Data scientist. DevOps engineer. Product manager as a distinct role. Customer success manager. Chief happiness officer — okay, that one is silly, but you get the point. Every major technology shift creates new categories of knowledge work, and those new categories are staffed by people who position themselves for them before the category has a name.

The best career of 2035 probably has a job title that has not been invented. That sounds like a problem. It is, in fact, really good news, and this chapter is about why, and how to be one of the people who gets there early.

— — —

Pattern Recognition, Not Title Prediction

Let me say up front what this chapter is not going to do. I am not going to give you a list of specific job titles and tell you the future belongs to "AI Alignment Coordinator" or "Synthetic Content Provenance Auditor." I could do that. Some of the titles I would give you might even turn out to be right. Most of them would not, because specific titles are rarely predictable, even when the underlying roles clearly are.

What I am going to do is describe the specific patterns that reliably show up in post-transition labor markets, so you can recognize the emerging categories in your own industry and position yourself for them before the job postings catch up.

The categories I have watched form over the last three years, across dozens of industries, cluster around five stable patterns. Every one of them will produce jobs in the 2030s that do not have names yet. Some of the jobs will be permanent. Some will be transitional. Either way, being early is the move.

— — —

Pattern One: The Hybrid Translator

When two previously separate domains are forced to work together at scale, the bottleneck is always the translator. The person who understands both vocabularies well enough to explain one to the other.

Examples that already exist: physician-informaticists who bridge clinical practice and health IT. Legal engineers who bridge law and software development. Quant researchers who bridge finance and statistics. Solutions architects who bridge customer problems and technical systems.

Examples emerging right now: AI-clinical integrators (people who help hospitals actually deploy AI clinical tools into real workflows). AI-legal integrators (people who design and supervise AI-assisted legal workflows at firms). AI-compliance integrators (people who translate between the AI engineering team and the compliance officers, and who keep both happy). AI-creative integrators (people who bridge traditional creative practice and agent-assisted production).

The hybrid translator role is never the most glamorous job in the room. It is often the most economically valuable one, because it is the person whose absence would cause the whole system to stop working.

To be a hybrid translator in your own field, you need two things. Deep enough fluency in each of the domains that neither side can dismiss you. And genuine respect for the practitioners in each domain, because translators who condescend to one side of the bridge are the ones who get fired first. The first thing takes years. The second thing is free and rare.

— — —

Pattern Two: The Governance Function

Every new powerful technology eventually produces a governance function. The role exists to decide what the organization will and will not do with the technology, and to be accountable for those decisions.

We already saw this play out with the internet. In 1995 there was no such thing as a Chief Information Security Officer. By 2005 every serious company had one. The role was created because the technology had created risks the organization could not ignore, and because a specific accountable human was needed to own those risks at an executive level.

AI is producing the equivalent wave right now. Chief AI Officer. Head of Responsible AI. AI Governance Committee Chair. AI Ethics Officer. Model Risk Officer. These titles did not exist five years ago. Some of them will not exist in their current form five years from now, but the underlying role — the accountable senior executive for AI risk and governance — is permanent. Every Fortune 500 company and every serious mid-market company is going to have this function. Most of them are still figuring out what the function looks like and who fits it.

If you can combine the technical literacy to understand what is being deployed, the regulatory and legal literacy to understand the rules, the organizational literacy to understand how to get things done inside a big company, and the temperament to say no to the CEO when that is the right answer, you are a candidate for this role at whatever level your career is currently at. The smaller the company, the earlier you can be the whole function. The bigger the company, the more the function is already staffed and you have to join a team rather than build one.

This pattern compounds the Compliance Native career from Chapter Twelve. The person who will end up running AI governance at their company in 2030 is often the person who, in 2026, spent a year quietly learning the EU AI Act and getting a privacy credential.

— — —

Pattern Three: The Supervisor of Automated Work

This is the category we have been circling for half the book — the agent wranglers, the fleet managers, the orchestrators. It is worth naming here because the specific titles are still emerging.

A few that are showing up in real job postings: Agent Operations Manager. AI Workflow Architect. Prompt Architect (note: this is different from Prompt Engineer; the architect designs systems). Automation Supervisor. Human-in-the-Loop Specialist. Model Quality Lead.

The underlying role is consistent across these titles. You are responsible for a portfolio of automated workflows that produce real business output. You design them, deploy them, supervise them, and take responsibility when they go wrong. You are a manager of non-human workers, and you are accountable for their output in the same way a traditional manager is accountable for a team's output.

This is the default role for a large share of displaced middle managers, if they have the literacy. Many of them do not yet. The ones who acquire the literacy in 2026 and 2027 are going to be the ones in these roles by 2028 and 2029, at compensation that is holding or rising.

— — —

Pattern Four: The Content Provenance and Trust Function

As AI-generated content fills the internet, the ability to verify what is real, what came from whom, what can be trusted, and what was produced by which system becomes economically valuable.

A few roles that are emerging in this space: Content Provenance Analyst. Authenticity Officer. Deepfake Forensics Specialist. Attribution Engineer. Trust and Safety Analyst (already exists at platforms; expanding). Synthetic Media Auditor.

Some of these titles will consolidate, some will change, but the underlying function will keep growing. Organizations that publish anything — media companies, political campaigns, financial institutions, educational institutions, courts — need to know whether the material they are relying on is authentic. The people who can verify this are going to be in strong demand. A journalist or investigator with forensic digital skills is the starter template for this career. The expansion from there into corporate, legal, and regulatory settings is happening fast.

— — —

Pattern Five: The Specialist Who Teaches Machines Their Niche

Every AI system that is good at something general needs to be made good at something specific. This specialization work is often called "fine-tuning" or "alignment for specific domains," and it requires the combination of deep domain expertise and basic AI literacy.

The emerging roles: Domain-Specific AI Trainer. RLHF Specialist (reinforcement learning from human feedback). Expert Reviewer for AI Model Development. Model Curator. AI-Assisted Content Specialist for specific verticals.

Who ends up doing this work? Typically not AI researchers. AI researchers want to work on the frontier, not on the specific applications. The people who do this work are domain experts — doctors, lawyers, teachers, writers, designers, financial analysts, compliance officers — who have learned enough AI literacy to participate meaningfully in the specialization process.

If you have deep expertise in any specific professional domain, and you add AI literacy on top of it, you are qualified for a category of work that almost nobody was qualified for three years ago, and that will keep expanding for the rest of the decade. This is the classic combination-chapter pattern applied to a very specific opportunity.

— — —

How to Position for Jobs That Do Not Exist Yet

Since I am not going to predict specific titles, let me give you five concrete behaviors that reliably position people for emerging roles.

98. **Be visible in the specific conversation.** Emerging roles get staffed from small, highly networked communities of people who are already talking about the problem. Find those communities in your field — Slack workspaces, LinkedIn subgroups, industry forums, specific conferences — and participate. Contribute. Be the person others recognize as thinking about this.
99. **Write or speak publicly about the problem.** Blog posts. Talks. Short videos. Podcast appearances. The act of publishing your thinking — even badly, even without a big audience — creates a breadcrumb trail that recruiters, hiring managers, and potential collaborators use to find the right hire for a role they may not have formalized yet.

100. **Take on adjacent projects in your current job.** Volunteer for the AI pilot. Write the governance memo nobody asked for. Build the prototype workflow. Get your fingerprints on the early work at your company, before the formal role exists. When the role is eventually defined, you are the obvious candidate.

101. **Build a portfolio of relevant artifacts.** For any emerging role, you can create example work that demonstrates you could do the role. Case studies. Workflows you designed. Governance documents you drafted. Research you conducted and published. Portfolio evidence is how you get hired for roles when you do not have the specific prior title to point to.

102. **Accept lower titles and unclear scope early, in exchange for first-mover position.** The people who end up running emerging role categories almost always started by doing the work under a title that did not fit, in a scope that was ambiguous, at a compensation level that did not match the eventual market rate. They gave up short-term clarity for long-term positioning. This trade-off is almost always worth it, provided you understand you are making it.

— — —

Diagnostic

- In the last six months, have I participated actively in any community or forum focused on an emerging problem in my field?

- Have I published any thinking — even a three-hundred-word post — about an AI-adjacent problem specific to my work?
- What adjacent project, at my current employer, could I volunteer for this quarter that would put my fingerprints on an emerging area?
- If someone were hiring for a role that did not have a title yet, and they needed to find me via Google or LinkedIn search, what search term would they use, and would I come up?
- Am I gripping a specific current job title too tightly to accept an unclear-but-strategic opportunity if it came along?

— — —

Monday Morning

Write a two-hundred-word LinkedIn post this week about an AI-adjacent problem specific to your field. Not a think piece. A concrete observation. Something you noticed. Something you are building. Something you think is being ignored.

Post it. It will probably not get much engagement. Do it anyway. Do one a month for the next year. At the end of the year, you will have a portfolio of twelve posts demonstrating that you think about emerging problems in your field, and your name will start showing up when recruiters and collaborators search for those topics.

That portfolio is how you get hired for a role that does not exist yet. Not tomorrow. Two or three years from now, when the role is formalizing and someone is trying to find the candidate who has been thinking about it the longest.

— — —

Science Anchor

- Historical BLS data on labor category creation. Look at how many current three-digit occupation codes did not exist in 1980. It is more than you expect.
- Research on labor market "churn" and category creation. Autor and others have written extensively about the pattern of new categories absorbing labor released from old ones, and about the conditions under which this does and does not happen smoothly.
- Stanford HAI reports on emerging AI-adjacent roles. Updated annually; useful for current-state snapshots.
- WEF Future of Jobs reports. Short on specific titles, useful on underlying categories.
- LinkedIn's annual Emerging Jobs reports. Flawed but directionally useful for seeing which titles are actually showing up in hiring data.

— — —

Close

The best career advice anyone can give a thirty-five-year-old in 2026 is almost certainly wrong in specifics and right in pattern. Nobody knows exactly what the premium jobs of 2035 will be called. Everybody who is paying attention knows roughly what they will look like.

If you optimize for the jobs that currently exist, you are optimizing for a snapshot of the labor market that is actively decaying. If you optimize for the underlying patterns — hybrid translation, governance, automated-work supervision, content trust, domain-specific AI specialization — you are optimizing for the structure that will produce the next decade's jobs, whatever they end up being called.

This is a subtle but important shift in how to run a career. Stop chasing titles. Start chasing patterns. The patterns are visible in 2026. The titles will follow.

The person who gets hired for the role that does not yet exist is almost never the person with the most polished resume in the category. It is the person who was early. The person whose name kept coming up when people talked about the underlying problem. The person who had already been doing the work, informally and for free, when the organization realized it needed to pay someone to do it formally.

That can be you. Start being early this week.

Next chapter: Part Five begins. The survival manual. We spend three chapters bringing everything together — the financial infrastructure you need to ride this out, the emotional and meaning work that matters when the old scripts stop applying, and finally a Monday-morning plan that takes every framework in this book and compresses it into specific actions you can take next week. Let us land this.

PART FIVE

The Survival Manual

Money. Meaning. Monday morning. The three pieces that turn a plan into a life.

Chapter 20: Money in a Post-Labor World

Let me start with a disclaimer, because this chapter is about money and I am not a financial advisor. Nothing in this chapter is specific financial advice for your situation. I am not licensed to give that. If you need specific guidance about your retirement accounts, your tax situation, your insurance choices, or your investment allocations, go talk to a credentialed fiduciary planner. What follows is conceptual framing for how to think about the financial side of the career transition we have been discussing. Please take it as framing, not instruction.

With that out of the way: we need to talk about money.

Nineteen chapters of this book have been about how to build a career that holds up through the cortex revolution. That is necessary but not sufficient. If you build a great career and you manage the financial side the way most Americans managed it in the 1990s, you are going to reach sixty-two with less optionality than you should have, and possibly a lot less. The old financial assumptions were calibrated to labor markets that no longer exist.

The good news is that the updates are not complicated. They are mostly things you already half-knew and had not gotten around to operationalizing. This chapter tries to close that gap.

A final note before we start. I am writing this at sixty-two. I am, right now, in the middle of exactly the decisions this chapter is about. I do not have it all figured out. I am not going to pretend I do. What I have is a concrete framework, tested against my own situation and against the situations of the people I have watched navigate this transition well or badly. Take what fits your life. Discard what does not.

— — —

What Changed, Specifically

The old playbook for middle-class American finance went something like this. Get a stable job with benefits. Stay for thirty years. Contribute steadily to a 401(k). Pay down the mortgage. Draw a pension or Social Security, combined with 401(k) withdrawals, starting in your mid-sixties. Health insurance through your employer until Medicare kicks in at sixty-five. Retire. Enjoy. Die quietly at seventy-six.

That playbook worked, with variations, for the Boomers and for much of Gen X. It assumed several things that are no longer reliably true.

- **Employment continuity.** The old playbook assumed you would mostly work for one employer, or a small number of employers, for decades. Current American workers under forty change jobs every two-to-three years on average. Career continuity is thinner. Benefits continuity is thinner. Retirement savings continuity is thinner.
- **Employer-provided benefits as a career constant.** Health insurance, retirement matching, life insurance, disability insurance — all of these were historically tied to W-2 employment with large employers. As the workforce has shifted toward contract, consulting, small-business, and gig work, these benefits have become less reliable and more expensive to replicate privately.

- **Pensions.** Fewer than one in ten private-sector workers has a traditional defined-benefit pension in 2026. The retirement risk has transferred from the employer to the employee over the last forty years. This transfer is complete. The 401(k) is now the retirement system.
- **Housing as a reliable appreciating asset.** Still mostly true, but less mechanically reliable than it was. Some markets have stalled. Some have reversed. Mortgage rates at the current level make the old "buy a bigger house every ten years" pattern much less viable than it was when rates were under four percent.
- **Linear career wage growth.** The old assumption was that you would earn more, in real terms, every decade of your working life, peaking in your fifties. Many current mid-career workers are finding that the wage-growth curve flattens or reverses earlier than they expected, especially if they are in AI-exposed knowledge work. Planning for continuous real wage growth is increasingly unsafe.

None of these changes are catastrophic individually. In combination, they require a different mental model than the one most of us inherited from our parents.

— — —

Optionality Is the Goal

The framing I have found most useful for thinking about money in this transition is simple. Money is not the goal. Optionality is the goal. Money buys optionality.

Optionality, in this context, means the ability to choose. To choose to stay in your current job or leave it. To choose to take a pay cut to learn a new field. To choose to say no to a bad offer. To choose to take care of a sick parent without losing the house. To choose to retrain for two years at fifty-five. To choose to retire five years earlier than planned when the medical issue arrives. To choose to keep working at seventy because the work is meaningful.

Every one of those choices requires financial cushion. The bigger the cushion, the more choices you have. The more choices you have, the less the labor market can compress you into a corner.

Americans who optimize for maximum income often end up with less optionality than they expected, because they spend proportional to the income and build high fixed costs that require the income to keep coming. Americans who optimize for optionality often end up with less income than they could have had, but more of the thing that matters — the ability to say no to a boss, a client, a situation, or a market.

The cortex revolution is going to hand out a lot of compressive pressure in the next decade. The people with the most optionality will navigate it well. The people with the least will be backed into decisions they did not want to make.

Build the cushion. Build the cushion. Build the cushion.

— — —

The Updated Financial Infrastructure

A concrete list of financial infrastructure that matters more, specifically, in the current transition. I am deliberately brief, because you should be reading a book by a real financial planner for the details, not this one.

- **A larger emergency fund than your parents had.** The traditional guidance was three-to-six months of expenses. For most knowledge workers in AI-exposed fields, twelve-to-eighteen months is the right ballpark in 2026. The cost of income disruption is higher than it used to be, because re-employment takes longer and often at lower compensation. A deeper emergency fund is the cheapest insurance policy you can buy against career turbulence.
- **Health insurance that works outside of W-2 employment.** Know what it would cost, what it would cover, and how you would acquire it in the gap. COBRA is bridge-only. ACA marketplace is the realistic option for most people. Health sharing ministries are not insurance and typically do not replace it. Having done this research in advance, before you need it, is optionality you can spend later.
- **Disability insurance.** Underused by knowledge workers, especially self-employed ones. A long-term disability before retirement age is the most common catastrophic financial risk most people face, bigger than premature death. If your employer provides it, read the policy carefully; many cover only own-occupation for a few years. If you are self-employed, get a private policy.
- **Retirement accounts that you own, not your employer.** Traditional IRAs. Roth IRAs. SEP-IRAs for self-employed income. Solo 401(k)s. These travel with you across job changes, career pivots, and gaps. Contribute early and consistently. The magic of the compounding math is not a joke; it really is that big.

- **HSAs if you are eligible.** Triple tax advantage, portable across employers, investable long-term. One of the most underused financial tools available to middle-class Americans.
- **Diversification beyond the single employer.** If your income comes from one source, your risk is concentrated. If you can add even modest secondary income streams — freelance work, rental income, dividend income, a side business, royalties — the risk profile of your household becomes materially more robust. This is why so many of the most financially resilient people you know have two or three things going at once.
- **Awareness of your fixed cost structure.** The ratio of fixed to variable costs in your household is the most important financial metric most people never calculate. A household with low fixed costs — modest housing, low debt service, minimal recurring subscriptions — can ride out six months of income loss without drama. A household with high fixed costs is one month from a crisis the entire time, without knowing it.

None of this is complicated. All of it is boring. The boring stuff is, as usual, what actually separates the people who navigate transitions well from the people who do not.

———

The Career-as-Portfolio Frame

One conceptual shift that has helped a lot of mid-career professionals I know is treating the career itself as a portfolio, not a monolithic identity.

In the old model, you had a career. Singular. You were a marketing executive, or a lawyer, or a software engineer. The career produced one paycheck, one set of benefits, one professional identity. If the career went sideways, everything else went with it.

In the new model, you have a portfolio of professional activities. Your primary income. Your secondary income. Your learning investments. Your portfolio of skills. Your network. Your reputation in your industry. Your intellectual property, if any. Each of these is a separate asset. Each can be developed independently. Each can absorb some shock without taking down the others.

A career portfolio in 2026 often looks like this. Primary W-2 job or consulting practice generating most of the income. A side project or secondary income stream generating ten-to-thirty percent of the total. A learning investment — a certification in progress, a new skill being developed — that is not yet paying but is building future optionality. A visible professional presence — writing, speaking, community participation — that is not monetized but is creating opportunities. A personal financial cushion big enough to support a pivot.

The reason this frame matters is that it changes how you think about decisions. When your career is monolithic, every decision is existential. When your career is a portfolio, most decisions are incremental. You can try things. You can kill things. You can rebalance. You are not gambling your whole identity on any single move.

The portfolio frame also gives you more resilience against AI-driven disruption. If one component of the portfolio gets displaced, the rest remain. The people who are thriving in this transition are, almost without exception, operating from a portfolio rather than from a monolith.

— — —

Retirement Is Different Now

A brief and incomplete treatment of retirement, because it deserves its own book and I am not the person to write it.

Retirement in 2026 is different from retirement in 1996 in several specific ways. People are living longer. Healthcare costs more. Pension reliability is lower. The Social Security outlook is solvent but constrained. Markets are unpredictable on any ten-year horizon. Employer-sponsored retirement matching is less generous than it was. Long-term care costs have risen dramatically faster than general inflation.

For most people under fifty-five reading this, the right planning assumption is probably that you will work, at least part-time, longer than you thought you would. Not because you cannot afford to stop, necessarily, but because the traditional sixty-five-and-done timeline is looking tight against the numbers. Many of the happiest retirees I know are working twenty hours a week at seventy-two, at something they love, in a field where experience is valued. That is not failed retirement. That is modern retirement.

For people over fifty-five, the math gets specific fast. Working one or two more years in your sixties is frequently worth more than a decade of aggressive investing in your forties, because the combination of continued earning, deferred withdrawals, and rising Social Security entitlement is extraordinarily powerful. If you are in a field you enjoy, the financial case for working longer often lines up with the quality-of-life case. If you are in a field you hate, the pivot into something you would enjoy doing for another decade may be the single highest-leverage decision available to you.

Whatever your situation, please get a credentialed fiduciary planner involved. The math of retirement is not something most of us are equipped to do on our own, and the cost of getting it wrong late is catastrophic. The cost of the planner, by comparison, is trivial.

— — —

Diagnostic

- If my primary income stopped tomorrow, how many months could I survive on current savings and assets at my current expenses? Be specific. Calculate it.
- Do I have health insurance I could maintain if I left my current employer? Do I know how much it would cost?
- Do I have disability insurance? Own-occupation or any-occupation? Does the coverage amount and duration fit my actual situation?
- What percentage of my current income comes from a single employer or client? Is that percentage higher than it should be?
- When did I last meet with a credentialed financial planner to review my situation? If the answer is "never" or "years ago," that is worth prioritizing in the next ninety days.

— — —

Monday Morning

Run the emergency fund number this week. Take fifteen minutes. Add up your checking, savings, and accessible non-retirement investments. Divide by your monthly expenses. That is the number. Write it down where you will see it.

If the number is under six months, everything else in your financial life is secondary to building it. Twelve months is the goal. This is not interesting. It is, however, the single most important financial move most readers of this book can make in the next year.

If the number is over twelve months already, look next at the career-portfolio frame. Is your income concentrated in one source? Is there a low-effort secondary stream you could establish? This is the next lever.

If you do not have a credentialed fiduciary planner, book an initial consultation with one this quarter. Ask for a flat-fee or fee-only engagement, not a commission-based one. The conversation will give you more clarity in an hour than a year of self-research.

—•—

Science Anchor

- Fidelity, Vanguard, and Morningstar research on retirement savings adequacy. Not marketing materials; actual research. Free online.
- Kitces.com (Michael Kitces) for technical planning content. Written for planners, accessible to careful readers, extremely high-quality.
- Bogleheads forum and wiki. Unglamorous, empirical, community-moderated. Where many of the most

financially competent Americans actually work out their plans.

- Research on the "three-legged stool" of retirement income (Social Security, retirement savings, and either pension or continued earnings). The empirical literature on what actually works.
- Your state's health insurance marketplace website. Basic, necessary, often ignored. Understand your state's ACA landscape before you need it.

— — —

Close

I am writing this chapter on a Saturday morning. My wife is doing bookkeeping for her business at the kitchen table. Our rough financial plan for the next fifteen years sits on my computer, updated monthly, adjusted quarterly, reviewed with a planner once a year. It is not fancy. It is not bulletproof. It is, however, a plan, which most Americans in our cohort do not have, and the existence of the plan is what converts the anxiety of this transition into something I can actually work with.

You will probably not retire the way your parents did. You will, on any current projection, live longer than they will. You will face healthcare costs they did not face. You will navigate a labor market that makes no promises. You will do all of this with less pension support and less employer loyalty than the prior generation had.

You will do fine, probably, if you start building the infrastructure this week. You will struggle if you wait another five years.

The cortex revolution is not going to kill the middle-class American financial life. It is going to require a different financial life, built for a different labor market. The updates are available. Most of them are free. The ones that cost money are cheap relative to the cost of not doing them.

Build the cushion. Build the portfolio. Build the plan. And, when you have done all of that, go back to the first nineteen chapters of this book and keep building the career, because the career is the engine, and the engine needs to keep running for longer than any of us were told it would.

Next chapter: meaning. Because a career plan without a reason behind it burns out inside three years, and we do not have three years to waste. If the old sources of meaning — the job title, the industry, the pension — are not going to carry us, we need new ones. And they are findable, and they are real.

Chapter 21: *Meaning When the 9-to-5 Dies*

Let me tell you the quiet part.

For most of my working life, I did not think about meaning. I thought about performance, deliverables, titles, promotions, the next credential, the next paycheck. I am not proud of this. It is true. I suspect, if you have been candid with yourself at all while reading this book, that it has been largely true for you too.

The reason most of us did not think about meaning for forty years is that we did not have to. The system handed us a framework. You went to school. You got a job. The job gave you an identity. The identity came with a community — your coworkers, your industry peers, your professional association. The identity came with a purpose — work hard, rise up, provide for your family, retire. The identity came with a story — the arc of a career. It was not a bad framework. It held up for four generations.

It is not holding up now.

Most of this book has been about what to do when the old labor rules stop applying. This chapter is about what to do when the old meaning rules stop applying. Because they do not apply either, and pretending they do is how people get fourteen chapters into their mid-life crises before they realize they have been trying to solve a spiritual problem with a career move.

This is the chapter where we put down the playbook for a moment and talk about the actual stakes.

— — —

The Meaning Problem

For American knowledge workers born between roughly 1950 and 1985, the default source of meaning was the career. This was true whether people acknowledged it or not. Our parents asked us what we wanted to be when we grew up, meaning what job we wanted. We picked one. We trained for it. We did it. The identity came bundled with the work, and for decades, the bundle held together well enough that most of us never had to unpack it.

When the career becomes unstable — through automation, through industry decline, through the kind of transition this book is about — the identity becomes unstable with it. The person does not just lose a job. They lose the delivery mechanism for their sense of self. That is not a human resources problem. That is a spiritual one, and no severance package solves it.

I have watched dozens of people go through this transition over the last three years. Some of them navigated it well. Some of them did not. The ones who navigated it well had one thing in common. Their identity was not fully downloaded into their job title. They had sources of meaning that survived the transition, because those sources had been running in parallel the whole time.

The ones who did not navigate it well had put everything into the career. Every ounce of identity. Every important relationship. Every sense of purpose. When the career wobbled, so did they, and they had nothing left to stand on while the ground under them shifted.

Let me be specific about where the durable sources of meaning actually are, because the good news is that they are all still available, and building them into your life is cheaper and easier than most of us assume.

— — —

Where Meaning Was Actually Hiding

Here is the secret the old framework obscured. The job was never the meaning. The job was the delivery mechanism.

What we actually wanted from our jobs — what we actually got, on the days that mattered — was one of three things. Contribution: the sense that our work affected someone for the better. Connection: the sense that we were in communion with other people who cared about something together. Craft: the sense of doing work well for its own sake, the pleasure of skill meeting challenge.

The old career framework bundled all three of these with a W-2 and a paycheck. We confused the bundle for the contents. We thought the meaning was the job.

It was not. It was the three things inside the bundle. And those three things — contribution, connection, craft — can be delivered through almost any life structure, if you are deliberate about building the structure. The W-2 is not a requirement. The title is not a requirement. The corporate apparatus is not a requirement.

This is not a trick. This is not an inspirational reframe. This is the actual structural observation. The meaning was in the three C's. It was not in the logo on your badge. If you can get the three C's delivered through a different structure than the one you used to have, the meaning will still be there.

— — —

Contribution, Connection, Craft

A brief word on each, because each requires deliberate attention in a life where the career no longer delivers them automatically.

Contribution is the sense that your work matters to someone. That somebody is better off because you showed up. This does not have to happen at a desk. It does not have to come with a salary. The volunteer who tutors a kid in algebra is contributing. The grandparent who teaches a grandchild to bake bread is contributing. The retired engineer who helps neighbors fix things is contributing. The hospice volunteer sitting with a dying stranger is contributing more than most salaried workers contribute in a quarter. If your paying work is contributing, great. If it is not, build other avenues that do. Contribution is a need, not a nicety. Humans without a sense of contribution get sick.

Connection is the sense of being in communion with people who care about something together. This happens at work. It also happens at many other places. Community groups. Faith communities. Running clubs. Book clubs. Mutual aid networks. The men's group that meets Thursday evenings. The quilting circle that meets Sundays. The neighbors who have dinner on each other's porches in the summer. These are not substitutes for a career. They are parallel sources of the same nutrient, and for people whose careers used to provide all their connection, they will have to carry more of the load now.

Craft is the pleasure of doing something well for its own sake. Skill meeting challenge. The moment the difficult thing comes together under your hands. This can happen at work. It can also happen in your woodshop, your kitchen, your garden, your sketchbook, your writing practice, your musical instrument. If your paying work has stopped offering craft, you can build craft elsewhere. Many of the most serene people I know in their sixties have a lifelong hobby that has nothing to do with how they earned a living, and that hobby is a substantial part of why they are serene.

The point is not to abandon your career. The point is to stop asking it to carry the whole load. The career may still be a legitimate source of all three C's. But it cannot be your only source, because if it ever stops being one of them — as it will, eventually, for most of us — you will not have a way to stand back up.

— — —

Grief Is Allowed

Before we go further I want to say something that most career books will not say.

When your career changes, it is okay to grieve.

For many of us, our careers are more than a paycheck. They are a decade or two or three of identity. They are the place where we made some of our closest friends. They are where we learned who we were in the world. When that changes — especially involuntarily — it is a real loss, and pretending it is not is how people make bad decisions trying to outrun feelings they have not named.

Grief is not weakness. Grief is the appropriate response to losing something that mattered. Give yourself thirty days. Sixty. Ninety. Talk to a therapist if it helps — and if you are in the cortex revolution, therapy is probably something to consider anyway, because the transition is harder than almost anyone admits. Talk to the people who know you. Do not try to power through.

The other side of grief is not bitterness. It is not denial. It is acceptance, followed by the work of building the next chapter. You cannot shortcut to the acceptance by skipping the grief. The grief is the shortcut.

I say this because I watched a man I respected burn through the first year of his post-career life in a rage that ate his marriage and his health. He refused to name what he had lost. He called every mention of it "pity," and he resented anyone who tried to offer it. Two years later, he was in a worse place than when he was laid off. Not because of the layoff. Because of the refusal to grieve.

You are allowed to feel this. You are, in fact, supposed to. The feeling is the first step toward being okay.

— — —

New Rituals for New Lives

One specific piece of structure that helps most people navigate this transition well: deliberate rituals.

The old career came with rituals built in. Monday morning coffee. The weekly staff meeting. The annual review. The office holiday party. The retirement lunch. These rituals were not trivial. They marked time. They connected you to other people. They gave shape to the week, the year, the career.

When those rituals disappear, most of us feel the absence as a vague disquiet we cannot name. We try to fill it with streaming services or doomscrolling or alcohol. None of those work. What feels like relaxation is actually numbing around a shape-loss that needs structure, not sedation.

What works is building new rituals, deliberately. Weekly. Monthly. Annually. Small gatherings with the same people on a predictable schedule. A Sunday walk with your partner. A monthly dinner with old friends. An annual weekend away with your college roommates. A Saturday morning in the workshop, in the garden, at the keyboard — whatever your craft is — on a schedule you do not move for inconvenience.

Pick some rituals. Make them non-negotiable. Show up to them the way you used to show up to the staff meeting. If you cannot show up to them the way you used to show up to the staff meeting, you are not serious about them yet, and you need to either raise the priority or pick different ones.

Rituals hold time together. Without them, life becomes a smear. With them, life has shape. This matters more than it sounds like it should.

One last note on rituals. Some of the best ones honor things that used to be inside your job. If you spent twenty years in a profession that valued reading and discussion, start a monthly book club. If you spent twenty years in a profession that required collaboration, join a community project that needs a team. If you spent twenty years doing work that you loved on some Tuesday mornings, find a way to keep those Tuesday mornings even if the work has changed. The ritual preserves the good part of the old life even after the structure of that life has shifted.

— — —

Diagnostic

- If my career ended tomorrow, what parts of my identity would survive?
- Do I have sources of contribution, connection, and craft that exist outside my job?
- Do I have at least three rituals, with specific people, on specific schedules, that hold my week and my year together?
- Am I grieving something I have not let myself name?
- When I imagine a version of my life twenty years from now that I would be proud of, does it require a specific career, or does it require a specific kind of person?

— — —

Monday Morning

This week, write down three sources of contribution, connection, and craft that exist in your life right now. Not aspirations. Real current practice.

If you cannot fill all three honestly, notice which one is empty. Make one small move this month to start filling it. A weekly ritual. A volunteer role. A hobby picked back up after a long absence.

The point is not a perfect life plan. The point is deliberate rather than drift. A deliberate life, aimed at contribution and connection and craft, will weather almost any labor market transition. A drifting life will not.

— — —

Science Anchor

- Viktor Frankl, *Man's Search for Meaning* (1946). The foundation text on meaning-making through work and suffering. Dated in some cultural particulars; timeless in its core argument.
- Robert Waldinger's Harvard Study of Adult Development, cited earlier. The longest-running study of human happiness in history. Its central finding — that the strongest predictor of late-life satisfaction is the quality of close relationships, not career prestige or income — is dispositive on the meaning question whether we want it to be or not.
- Robert Putnam, *Bowling Alone* (2000) and his later work. The civic and associational fabric of American life has thinned, not thickened, since 1970. This is part of why meaning feels harder now. Know the terrain you are in.
- Research on the Japanese concept of ikigai, particularly academic treatments rather than the pop-culture four-circle graphic. The underlying idea — that meaning in work sits at the intersection of love, skill, need, and sustenance — is more robust than the simplified version suggests.
- Research on the health effects of loneliness and low social integration. The body reacts to an absence of connection the way it reacts to smoking. This is not a metaphor. This is physiology.

— — —

Close

I am writing this chapter the night before my sixty-second birthday.

I am tired today. Not bad tired. Just the ordinary tired of a person who has done a full week of work and is at the age where the week leaves a mark. My wife is reading in the other room. The cat is asleep on the sofa. There is cold coffee on my desk and a half-eaten sandwich next to my keyboard, and I am about four hundred words from finishing this chapter, and tomorrow there is a birthday, and after that another day, and after that another.

None of this is exciting. None of it will be in an obituary. All of it is what a life is actually made of.

I used to think meaning was something you achieved. A big goal, accomplished. A monument in whatever field you were in. A thing that would be remembered. I do not believe this anymore. Meaning is what you find in the actual hours, with the actual people, doing the actual work, in a body that is slowly, inexorably aging, in a life that is finite.

The cortex revolution is going to take a lot of the old structures that we used to wrap around our lives. Careers. Titles. Pensions. Predictable trajectories. It is going to ask a lot of us in the way of adaptation. I hope you are up for the adaptation. I believe, based on watching people navigate it for three years, that most of us are.

What it cannot take is the three C's. Contribution. Connection. Craft. And it cannot take the quiet Tuesday evenings. It cannot take the ritual. It cannot take the birthday, and the dog, and the coffee, and the half-eaten sandwich.

Those are yours. If you hold them — gently, deliberately, over decades — and build your life around them rather than around a job title, you will be okay. Whatever happens to your specific career. Whatever happens to the broader economy. You will be okay.

That is the point. That has always been the point.

Next chapter: the last one. The Monday morning plan. All five characters, one last time. And then you and me, at the end of a book together, at the start of a week that is yours to make.

Chapter 22: The Monday Morning Plan

You have read twenty-one chapters. Now there is a Monday morning. What do you do?

This is the last chapter. It is also the most practical one in the book. For two hundred-odd pages I have been building a framework. Here is where it compresses into specific actions you can take next week. Not next quarter. Not after you feel ready. Next week.

I will also, in this chapter, check in one final time with the five people who have walked through the book alongside us. Dave in Cleveland. Marisol in Phoenix. Jamal in Austin. Dr. Eleanor Voss in Boston. Kevin in the Midwest. You will see what happened to each of them. You will see, in each case, that the outcome was a function of specific decisions made at specific points, not of luck or talent or timing. You are at one of those points right now. What you do in the next thirty days will matter more than what you did in the last three years.

Let us land this.

— — —

The Monday Morning Seven

The entire argument of this book, compressed into seven actions. Do them in rough sequence. Do not try to do all seven in one weekend. One per week for the next seven weeks is aggressive but realistic. Slower is also fine, as long as you are moving.

One: The Job Inventory. Open a document. Call it "My Job, Honestly." List every recurring task you do in a week. Next to each, write A (could be done by an AI tool for under a hundred dollars a month), B (requires you, but only due to institutional inertia, and will be A within eighteen months), or C (genuinely requires human body, human signature, or human judgment in a messy situation). This is the diagnostic from Chapter One. Almost nobody actually does it. Do it.

Two: The Pillar Audit. Look at your C tasks. Which of the three pillars do they rest on? Physicality (your body, in a specific place, doing something unpredictable). Accountability (your license, your signature, your personal liability). Taste (your judgment, your curation, your ability to tell good from great). Most durable careers rest on one or two of these pillars. Name yours. If none of your current tasks rest on any of them, that is the signal, and you already know what it means.

Three: The Pivot Target. Given your pillars, your skills, your credentials, and your life circumstances, name one specific role you could plausibly hold in twelve to eighteen months. Not an aspiration. A concrete target. A Licensed Signer role. A physical-plus-judgment role. A taste-forward curation role. An agent-wrangler role. A compliance-native role. A clinical healthcare role. A solo-operator role. Pick one.

Four: The Credential Path. What certification, license, or degree does the pivot target require? Find out. Specifically. This week. Look up the program. Call the admissions office. Understand the timeline, the cost, and the sequence. Do not skip this step. The credential is usually the cheapest part of the pivot, and people who skip it often end up unhireable for their own target role five years later.

Five: The Agent Workflow. Build one multi-agent AI workflow for a recurring task in your current job this month. Not a single prompt. A workflow. Three agents, minimum, handing off to each other. Ship it. Use it weekly. Refine it. This is your proof to yourself and to future employers that you have crossed from Level One to Level Two of the new literacy.

Six: The Financial Floor. Calculate your current emergency fund in months of expenses. If it is under six months, make building it your top financial priority for the year. Twelve to eighteen months is the target for anyone in AI-exposed knowledge work. Optionality is the goal. The fund is what buys it.

Seven: The Three C's. List your current sources of contribution, connection, and craft. If any is empty, make one move this month to begin filling it. Rituals help. Deliberate structure helps. Drift does not help. A life organized around the three C's survives any labor market. A life organized around a job title survives exactly that job title.

That is the book. Seven moves. Start this week. Finish in two months. By the end of this year, you will have moved more decisively on your own career than ninety-five percent of the people in your industry. The payoff is cumulative, and the timeline is shorter than you think.

— — —

Dave, Nineteen Months Later

Dave Palmieri is forty-nine now. He is still in commercial trucking underwriting. His title has changed. He is now the Director of Commercial Risk Accountability at his firm, a role the firm created specifically for him after the CFO realized that somebody needed to actually own the sign-off on AI-generated underwriting decisions, and that Dave was the only person in the office who both understood the book of business and was willing to take the liability seriously.

He manages a team of three other licensed underwriters and oversees roughly two hundred AI-produced underwriting recommendations per day. His compensation is up about thirty percent from 2024. His firm has begun citing him as a model for the new shape of the role at industry conferences. Two smaller competitors have tried to poach him, and he has turned them down, because his current firm has given him the seat at the table he needed.

He still thinks about the nine people who did not make it. He stays in touch with most of them. One of them — the man who called him the Saturday after the severance — has since completed a certified financial planner credential and is building a practice of his own focused on mid-career professionals navigating exactly this kind of transition. Dave sends him referrals, and the man sends Dave cigars at Christmas.

Dave, in short, is not thriving because the cortex revolution was kind to him. He is thriving because he saw what it required, made a specific pivot, and did the work. The same move was available to every member of his old team. The ones who made it are okay. The ones who waited are still waiting.

— — —

Marisol, a Year After the Decision

Marisol Quintero is thirty-five now. She decided, after three months of running the numbers and visiting the competitive landscape, to open the Chandler location. The Chandler location opened six months ago. It is already breaking even on a cash basis, which is faster than her own accountant projected.

She now has twenty-two employees across the two locations. She has hired her first formal apprentice through a women-in-the-trades program in the Phoenix metro. She has begun negotiating with a local community college about sponsoring an HVAC apprenticeship track, because the labor shortage is now her single biggest operational constraint, and she has decided that if the pipeline is not going to fix itself she will fix a small piece of it directly.

She has also purchased a small parcel of commercial land in a growing suburb east of Phoenix, with an eye toward either a second business concept — she has been quietly researching commercial refrigeration services, which have a worse labor shortage than residential plumbing — or eventually retiring onto it in fifteen years. She has not decided yet.

When she heard that this book was almost finished, she asked if she was going to be in it. I told her yes, and that she had opened Chapter Fifteen. She laughed and asked if the cortex revolution guys in the book had figured out how to crawl under the 1962 bungalow yet. The answer continues to be no. Marisol is not worried. Marisol has never been worried. Marisol is, in fact, the single most economically secure person in this book, and she arrived at that security by taking a career path most American parents still apologize for at dinner parties.

— — —

Jamal, at Year Three of the Bankole and Co. Experiment

Jamal Bankole is thirty now. Bankole and Co. has grown — deliberately, slowly — to three human employees plus his agent fleet, which is now at roughly sixty workflows. Revenue is up to around six million. Margins are still roughly forty percent.

He has turned down two acquisition offers from larger agencies that wanted to absorb his operation and his methodology. He is not interested. He has begun a parallel practice — quarterly workshops for mid-career marketers pivoting into agent-assisted work — that is now roughly a quarter of his revenue and almost all of his teaching time.

He has become a recognizable voice in his industry. He has also become increasingly aware that the operating model he pioneered is becoming less exotic every quarter, as his students build their own versions of it and his competitors either adapt or shrink. He is okay with this. He did not build the model to keep it scarce; he built it to do better work and have a better life. That work continues. The life continues. His fiancée, for the record, married him last spring.

When I asked him what he wanted mid-career readers of this book to understand, he said: "Tell them the hardest part is the first ninety days. Tell them to push through it. Tell them that after ninety days, it gets easier every week for the rest of their career. Most people quit at day sixty. The ones who do not quit end up where I am. That is the whole secret."

That is the whole secret.

— — —

Eleanor, on the Question of Retiring

Dr. Eleanor Voss is sixty-four now. She has not retired, will not retire at sixty-five, and is no longer asked about it. She is, however, restructuring her practice in a way that will let her reduce direct patient care over the next three years while expanding her supervision and consultation with younger clinicians.

She is now supervising twelve early-career clinicians across the Boston area. She sees eighteen of her own patients weekly, down from twenty-two. The supervision work, in her words, "is the most important thing I have ever done professionally." She is helping shape the next generation of clinicians who will inherit the workforce shortage when she finally steps back, if she ever fully does.

She has begun writing. A book. She will not say yet what it is about. She will only say that it is not a memoir, and that it is the kind of book that is best written slowly, in her sixties, by someone who has spent forty years in the consulting room. Her publisher — she has one — is being patient.

When I asked her how she thought about meaning at this stage of her career, she thought for a long moment, as she always does, and then said: "Meaning is what I am carrying for people now. I carry some of what they cannot carry yet. When they can carry it themselves, they stop needing me. That is the whole job. That is what I get paid for. That is what I will always have been for."

I wrote that down. I am writing it down for you here, too, because I think it generalizes to a lot of us, in a lot of fields, more than we admit.

— — —

Kevin, Two Years Out of College

Kevin Huang is twenty-four now. He graduated, as promised, with the useless business degree. His parents were proud. He hugged his mother at the ceremony and did not cry until he got to the car.

Kevin did not, after graduation, take the first marketing job he was offered. That was the single most important decision he has made so far, and he has told me he almost took the job out of exhaustion and the need to start paying his loans. Instead he took a lower-paying customer-research role at a mid-market brand agency, specifically because the role put him in contact with senior researchers who were starting to use AI tools in ways that fascinated him.

Inside eighteen months, he had identified — using the pattern framework from Chapter Nineteen — that content provenance analysis was going to be a real role within three years, and had taught himself the specific skills he thought it would need. He had been writing publicly about the space. He had a small portfolio of case studies. He had become, quietly, one of the small number of people in his local market who had been thinking seriously about the problem.

Six months ago, a mid-market PR firm decided it needed someone to own content provenance for its client work and began asking around about who was thinking about the problem. Three separate people pointed them at Kevin. He is now the firm's first-ever Content Provenance and Authenticity Analyst, at a compensation package that exceeds what his old senior marketing boss was making at forty-five.

He is twenty-four.

When I asked him what he wished he had known two years ago, he said: "I wish I had known it was going to be okay. I was so scared that whole senior year. I did not know how to tell my parents that the thing they paid for was a bad bet. I did not know there was a path out. If I had known even one person who had made it through, I think I would have slept better." He paused. "Tell other kids. Tell them it is going to be okay."

Tell other kids. Tell them it is going to be okay.

— — —

A Personal Note

I am finishing this book on a Monday afternoon in Florida. It is hotter than it should be this early in the year. My wife has a job on the other side of town and will not be home until six. The cat is asleep, again. I have written about sixty thousand words across twenty-two chapters, and I am about two hundred words from being done.

I do not know, honestly, whether I have said the right things. I hope so. I have tried to be truthful about the labor transition I see unfolding, truthful about the human premium that remains, truthful about what I do not know, and truthful about the emotional weight of writing a book at sixty-two that is largely about the advice I wish somebody had given me at forty-five.

I have lived some of what is in this book. The voice recording was real. The cervicogenic migraine was real. The sixty-third birthday is tomorrow. I am taking some of my own advice, quietly, in the evenings — the Friday unlearning hour, the agent wrangling, the three C's audit. Some of it is working. Some of it I am still figuring out. I am not ahead of you. I am in this with you.

Thank you for reading to the end. If you made it here from the beginning, you are probably the kind of person who does not drift. You are a deliberate person in a moment that rewards deliberateness. That is in your favor. That is, actually, most of the game.

— — —

Close

A long time ago somebody told me that the point of a good book was not to change your life. It was to give you one useful frame you did not have before, and let you carry that frame with you into the rest of your life.

I hope this book gave you at least one. The three pillars. The accountability premium. The starfish shape. The three C's. Any one of them, held and used, is worth the cost of the book.

The cortex revolution is going to be the defining labor transition of your working life and mine. We are going to live through it. Most of us are going to do fine, eventually, on the other side. The ones who do fine will have been deliberate early, started small, kept their cushions, honored their relationships, and refused to tie their identities too tightly to any single professional structure.

You can be one of them. You are, in fact, better positioned than most to be one of them, because you have read a book that almost nobody else has read, and you know things about how this is going to unfold that most of your neighbors do not yet know.

Start this week. Take one action from the Monday Morning Seven. Then another next week. Then another. Tell one person what you are doing. Ask them to check in on you in three months. Give yourself permission to adjust as you go.

You are going to be okay.

You are going to be useful, to people who matter to you, for a long time.

You are, to use the title of this book one last time, still useful — and you will remain so, through whatever the next twenty years bring. I am rooting for you. So are Dave, and Marisol, and Jamal, and Eleanor, and Kevin, and the rest of the cast of characters who do not have names in this book but whose lives are shaped by every sentence I have tried to get right.

Go get Monday.

— Ken Konet, Florida, April 2026.

APPENDICES

Tools, frameworks, and resources to move from reading to doing.

Appendix A: *90-Day Human Hardening Plan*

Twelve weeks. One chapter's worth of actions per week. Each week has a single goal, two to four specific actions, and a reflection prompt. At the end of ninety days, you will have executed the core moves from every chapter of this book in the right sequence, and the ongoing rituals will run themselves for the rest of your career.

Do not try to do this in thirty days. Aggressive compression produces burnout and skipped steps. Do not let it stretch to six months either; the gap between intention and action widens every week you delay.

———

Month One: Diagnose

Week 1: Assessment. Goal: know where you stand. Actions: (1) Complete the "My Job, Honestly" A/B/C inventory from Chapter One. (2) Complete the Pillar Audit — which of the three pillars (Physicality, Accountability, Taste) does your current work rest on? (3) Take the 25-Question Self-Assessment in Appendix F. Reflection: what did you learn that you did not already know?

Week 2: Pivot Target. Goal: name the destination. Actions: (1) Identify one specific role you could plausibly hold in 12–18 months. (2) List the credentials, skills, and portfolio that role requires. (3) Research the on-ramp specifically — program, cost, timeline. (4) Call one person who holds that role and ask three questions. Reflection: how plausible is this target, honestly?

Week 3: Install the Rituals. Goal: build the practice. Actions: (1) Put a weekly Friday Unlearning Hour on your calendar, recurring, non-negotiable. (2) Put a weekly 3-hour Cross-Domain Input block on your calendar. (3) Choose your first Scary New Thing for the coming quarter. (4) Start the 30-day Taste Training Protocol from Chapter Six. Reflection: which ritual feels easiest, which feels hardest, why?

Week 4: Financial Floor. Goal: build optionality. Actions: (1) Calculate your current emergency fund in months of expenses. (2) Set a 12-month target and a monthly contribution plan. (3) Review your fixed-cost structure and identify one reduction. (4) Book an initial consultation with a credentialed fee-only financial planner. Reflection: how many real months of optionality do you have today?

— — —

Month Two: Build

Week 5: First Agent Workflow. Goal: cross into Level Two literacy. Actions: (1) Identify one recurring task that takes 30+ minutes weekly and has structured inputs/outputs. (2) Build your first three-agent workflow for it. (3) Ship it and use it this week. (4) Document what worked, what failed, and what surprised you. Reflection: what does it feel like to supervise an agent versus prompt a chatbot?

Week 6: The Credential Move. Goal: start the paperwork. Actions: (1) Enroll in a course or order prep materials for your chosen credential. (2) Schedule the exam window 6–12 months out. (3) Find a study partner or community. (4) Commit publicly — tell at least one person you are doing this. Reflection: why did I wait?

Week 7: The Portfolio Piece. Goal: build visibility. Actions: (1) Write and publish one 300-word post related to your pivot target. (2) Identify the community where practitioners in that field talk — Slack, LinkedIn, industry forum, etc. (3) Join it and participate twice this week. (4) Identify one senior person in the field and send a thoughtful message, not a cold pitch. Reflection: who would find me via search today if they were hiring for this role?

Week 8: The Three C's. Goal: check the life structure. Actions: (1) List your current sources of Contribution, Connection, and Craft. (2) Identify which C is empty or thin. (3) Plan one monthly ritual that addresses it. (4) Begin that ritual this month. Reflection: which C was hardest to fill honestly?

— — —

Month Three: Expand

Week 9: Second Agent Workflow. Goal: demonstrate the pattern. Actions: (1) Build a second, more ambitious workflow using lessons from the first. (2) Connect it to a real business process. (3) Measure output against the manual baseline — time, quality, throughput. (4) Share the case study internally or on a public channel. Reflection: can I now describe agent orchestration in three sentences?

Week 10: The Network. Goal: get known. Actions: (1) Reach out to five people in or adjacent to your target field. (2) Have two or three actual conversations this week — coffee, Zoom, phone. (3) Ask each: "what does someone like me need to know?" (4) Follow up with something useful to them. Reflection: what pattern am I hearing across conversations?

Week 11: The Checkpoint. Goal: honest review. Actions: (1) Reread your Week 1 assessments. (2) Compare where you thought you would be versus where you are. (3) Update the pivot target if needed — small adjustments are healthy; large changes usually mean you picked the wrong target. (4) Plan Months Four through Six. Reflection: what is the single most valuable thing I have learned in eleven weeks?

Week 12: The Shift. Goal: make the new pattern permanent. Actions: (1) Confirm all recurring rituals are on the calendar indefinitely. (2) Schedule the Quarter Two Scary New Thing. (3) Confirm your credential exam is scheduled. (4) Tell one person what you are doing and ask them to check in with you quarterly. Reflection: who am I now that I was not ninety days ago?

— — —

Close

You have done the first ninety days. The momentum is now doing most of the work. Most of the rituals you installed will run for the rest of your career, because you built them into your calendar rather than into your willpower. The credential is scheduled. The portfolio exists. The network is warming up. The emergency fund is growing.

You are not finished. You will never be finished. But you have crossed the line from intention to motion, and that is the line most people in your position never cross. Keep going.

Appendix B: *Industry Survival Matrix*

Forty professions, each scored on four dimensions. AI Exposure (1–5; higher means more of the job can be automated with 2026 tools). Physicality (1–5; higher means more physical, in-person, unpredictable-environment work). Accountability (1–5; higher means more regulatory licensure and personal liability). Taste (1–5; higher means judgment and curation are the core value delivered).

These scores are directional, not precise. They are designed to trigger useful thinking, not to settle debates. A role with low total score in the three positive pillars and high AI Exposure is in the blast radius. A role that stacks high scores across multiple positive pillars against modest AI Exposure is a durable career in the cortex revolution.

Use this matrix to check your current role and your candidate pivot targets.

— — —

White-Collar Knowledge Work

1. Paralegal (junior). Exposure 5 / Physicality 1 / Accountability 2 / Taste 2. The middle of the legal stack is hollowing out first. Pivot to senior paralegal, compliance, or a licensed pathway.

2. Paralegal (senior, specialized). Exposure 4 / Physicality 1 / Accountability 3 / Taste 3. Some survival, especially in specialized niches. Long-term, move toward signer-adjacent work.

3. Corporate attorney (senior, signing). Exposure 3 / Physicality 1 / Accountability 5 / Taste 4. Signer role. Stable and rising in value as AI expands junior-work throughput.

4. Bookkeeper (non-credentialed). Exposure 5 / Physicality 1 / Accountability 2 / Taste 1. Deeply exposed. Pivot to CPA credential or controller-level roles.

5. CPA (senior, signing). Exposure 3 / Physicality 1 / Accountability 5 / Taste 3. Signer role. Demand for licensed tax and audit signoffs rising.

6. Marketing coordinator. Exposure 5 / Physicality 1 / Accountability 1 / Taste 2. Deeply exposed. Move toward strategy, analytics, or brand direction.

7. Marketing director (strategic). Exposure 3 / Physicality 1 / Accountability 2 / Taste 5. Taste-forward role. Durable if you are genuinely the decision-maker, not the coordinator.

8. Customer service representative (scripted). Exposure 5 / Physicality 1 / Accountability 1 / Taste 1. Among the most exposed roles in the economy. Pivot direction is escalation-management or specialized support.

9. Financial analyst (junior). Exposure 5 / Physicality 1 / Accountability 2 / Taste 2. Ladder rung being removed. Move toward licensure (CFA, CPA, Series) or client-facing advisory.

10. Financial advisor (CFP, fiduciary). Exposure 2 / Physicality 1 / Accountability 5 / Taste 4. Strong survival. Presence premium stacks on signer role.

11. Software developer (junior, boilerplate). Exposure 4 / Physicality 1 / Accountability 2 / Taste 2. Being compressed. Move toward specialty, senior, or orchestration roles.

12. Software developer (senior, specialist). Exposure 3 / Physicality 1 / Accountability 3 / Taste 4. Durable at senior levels. Specialty and system design protect.

13. Product manager (senior). Exposure 3 / Physicality 1 / Accountability 3 / Taste 4. Taste matters. Strategic PMs more durable than coordinating PMs.

14. UX designer (senior). Exposure 3 / Physicality 1 / Accountability 2 / Taste 5. Taste-forward. Durable if you can genuinely direct rather than execute.

15. Data analyst (general). Exposure 5 / Physicality 1 / Accountability 2 / Taste 3. Exposed. Move toward data science, domain-specialist analytics, or model governance.

16. Middle manager (knowledge work, coordinating). Exposure 5 / Physicality 1 / Accountability 2 / Taste 2. Among the most exposed roles. Pivot to agent wrangler, function owner, or solo-operator-inside.

17. Administrative assistant. Exposure 5 / Physicality 1 / Accountability 1 / Taste 2. Long-running decline, accelerated. Move toward specialized executive assistant or operations roles.

18. Content writer (junior). Exposure 5 / Physicality 1 / Accountability 1 / Taste 2. Deeply exposed. Move to taste-forward editor or domain-specialist roles.

19. Technical writer (senior). Exposure 4 / Physicality 1 / Accountability 2 / Taste 4. Partially protected by taste. Durable in specialized technical domains.

20. HR coordinator. Exposure 5 / Physicality 1 / Accountability 2 / Taste 1. Exposed. Pivot toward HR business partner, compensation specialist, or labor-relations.

— —

Healthcare

21. Primary care physician. Exposure 2 / Physicality 4 / Accountability 5 / Taste 4. Premium career. Structural shortage, signer role, presence required.

22. Registered nurse. Exposure 2 / Physicality 5 / Accountability 5 / Taste 4. Bulletproof. Every pillar stacks.

23. Physical therapist. Exposure 2 / Physicality 5 / Accountability 5 / Taste 4. Bulletproof. Aging-driven demand is structural.

24. Radiologic technologist / MRI / Ultrasound. Exposure 3 / Physicality 5 / Accountability 4 / Taste 3. Highly protected. Short training, strong wages, chronic shortage.

25. Psychiatrist. Exposure 2 / Physicality 3 / Accountability 5 / Taste 5. Premium career. Among the most durable roles in medicine.

26. LCSW / LMFT / LPC (licensed therapist). Exposure 2 / Physicality 3 / Accountability 5 / Taste 5. Bulletproof. Structural shortage; presence required.

27. Pharmacist. Exposure 3 / Physicality 3 / Accountability 5 / Taste 3. Strong. Specialty and hospital pharmacy especially durable.

28. Home health aide. Exposure 1 / Physicality 5 / Accountability 2 / Taste 2. Physically protected. Demand wildly outstripping supply; wages rising.

29. CNA (certified nursing assistant). Exposure 1 / Physicality 5 / Accountability 3 / Taste 2. Physically protected. Short training; gateway to broader nursing careers.

— — —

Skilled Trades

30. Plumber (licensed). Exposure 1 / Physicality 5 / Accountability 4 / Taste 3. Premium career. Physicality Wall plus accountability premium plus small-business ownership.

31. Electrician (licensed). Exposure 1 / Physicality 5 / Accountability 5 / Taste 3. Premium career. All four pillars favor.

32. HVAC technician (licensed). Exposure 1 / Physicality 5 / Accountability 4 / Taste 3. Premium career. Demand curve matches the trades shortage.

33. Welder (specialty: pipeline, aerospace, underwater). Exposure 1 / Physicality 5 / Accountability 4 / Taste 4. Premium career. Specialty welders are among the best-paid tradespeople in America.

34. Carpenter. Exposure 1 / Physicality 5 / Accountability 3 / Taste 4. Strong. Finish carpentry taste-forward; framing commodity-tier but protected by physicality.

———

Other Professions

35. Teacher (K–12). Exposure 2 / Physicality 4 / Accountability 4 / Taste 4. Presence-protected. Durable, especially in specialty teaching.

36. Compliance officer. Exposure 2 / Physicality 1 / Accountability 5 / Taste 4. Bulletproof. Regulatory tailwind plus signer role.

37. Real estate agent (established practice). Exposure 3 / Physicality 3 / Accountability 4 / Taste 4. Strong for established agents; hostile environment for new entrants.

38. Chef (mid-level professional kitchen). Exposure 2 / Physicality 5 / Accountability 3 / Taste 5. Strong. Taste + physicality stack.

39. Credentialed coach (ICF PCC or above). Exposure 2 / Physicality 3 / Accountability 3 / Taste 5. Strong. Taste + presence; accountability thinner due to lack of state licensure.

40. Peer support specialist / community health worker. Exposure 1 / Physicality 3 / Accountability 3 / Taste 3. Growing field; presence and lived experience are protective.

———

How to Use the Matrix

If your current role scores poorly against three of the positive pillars and has high AI Exposure, you are in the blast radius. Start the 90-Day Plan in Appendix A this week.

If your current role stacks 4+ on at least two positive pillars, you are in a durable position — focus on compounding it by adding agent literacy and maintaining your credentials.

If you are considering a pivot, score the target role on all four dimensions before committing. A pivot from a 4–4–4 to a 1–5–5 may look like downgrade on paper; in durability terms, it is an upgrade.

Scores reflect the labor market of 2026. Some may shift. Directional logic will continue to apply for the rest of the decade.

Appendix C: *Prompt Architecture Starter Kit*

Five concrete multi-agent workflow templates. Each follows the same pattern. A defined goal. A sequence of specialized agents handing off to each other. Clear inputs and outputs at each stage. Human checkpoints at the moments that require judgment or accountability.

These are templates, not recipes. Adapt the specifics to your tools, your domain, and your data sources. The point is the structure. If you build and ship even one of these, you have crossed from Level One to Level Two fluency, and the pattern will generalize to the rest of your work.

— — —

Template 1: The Research Brief Workflow

Goal. Turn raw client materials plus external research into a client-ready brief, reducing total human time from a full day to roughly forty minutes.

Agents. (1) The Ingestor takes raw client materials — emails, documents, notes, transcripts — and produces a structured summary of client priorities, stated constraints, and implicit open questions. (2) The Researcher takes those open questions and conducts external research, producing a research memo with sources and confidence levels. (3) The Synthesizer takes outputs from Ingestor and Researcher and produces a draft synthesis of where the client context intersects with the research findings. (4) The Drafter takes the synthesis and produces the written brief in your house style and formatting conventions. (5) The Checker takes the draft and runs it against a self-written checklist — claim-source alignment, tone appropriateness, compliance flags, missing caveats.

Human checkpoints. Initial scope approval before Ingestor runs. Final review before brief goes to client.

Variations. Swap "research brief" for "investment memo," "case summary," "marketing brief," or "policy analysis." The five-agent pattern generalizes across any knowledge-work deliverable that combines client context with external research.

— — —

Template 2: The Client Communication Triage Workflow

Goal. Manage inbound client communications at scale without dropping important threads or spending hours drafting routine replies.

Agents. (1) The Classifier reads incoming messages and sorts them into categories: urgent, routine, informational, escalation. (2) The Context Puller fetches relevant client history, prior communications, and open matters for each message. (3) The Drafter produces a response draft in your voice using context and category-specific templates. (4) The Tone Checker reviews drafts for sensitivity, boundary, and compliance flags.

Human checkpoints. Review drafts before sending; handle urgent and escalation categories personally.

Variations. This template works for client-facing practitioners in law, accounting, financial advice, consulting, or any relationship-heavy service business. The core insight is that the bulk of the time spent in client communication is routine and delegable; the judgment-required slice is small and should be the only part that takes your attention.

— — —

Template 3: The Content Review and Feedback Workflow

Goal. Review AI-generated or human-produced content at volume, catching quality failures and brand drift without manual review of every artifact.

Agents. (1) The Quality Screener reads each piece of content and sorts it into three tiers: reject, needs work, acceptable. (2) The Brand Alignment Checker scores content against your documented brand voice, producing a brand alignment score and specific drift flags. (3) The Fact Checker verifies factual claims against source materials and flags unverified or suspect assertions. (4) The Feedback Drafter combines outputs from the prior three agents into actionable revision notes for the content creator.

Human checkpoints. Borderline cases between "needs work" and "acceptable." All rejections before sending feedback. High-stakes content before publication regardless of scores.

Variations. Works for blog posts, social content, internal documents, technical writing, marketing collateral, educational materials. Particularly valuable when you are managing a team of AI-assisted content producers and need to maintain quality control without becoming the bottleneck yourself.

— — —

Template 4: The Weekly Industry Intelligence Workflow

Goal. Stay current in your field without spending hours scanning headlines, receiving a weekly digest of what actually matters instead.

Agents. (1) The Scanner pulls articles from your chosen sources — publications, newsletters, preprint servers, industry forums — over the past week. (2) The Summarizer produces concise summaries of each article with source links. (3) The Relevance Ranker scores summaries against your stated priorities and interests, producing a ranked list. (4) The Digest Builder assembles the top items into a Monday-morning digest, formatted for quick consumption.

Human checkpoints. Review digest Monday morning; decide which items warrant reading the full article.

Variations. This is the template I use personally to track developments in AI, labor economics, publishing, and a handful of adjacent fields. The setup takes a weekend. The time savings are enormous after the first month, as the agent learns your actual priorities from what you do and do not click through to.

— — —

Template 5: The Personal Finance Review Workflow

Goal. Monthly personal finance review without opening a spreadsheet, producing a one-page overview of where your money actually went and what flags need attention.

Agents. (1) The Transaction Categorizer ingests monthly statements and categorizes every transaction using your personal category scheme. (2) The Budget Comparator compares actual spending against your targets and produces a variance report. (3) The Flag Generator highlights unusual transactions, large variances, or categories trending in concerning directions. (4) The Insight Summarizer rolls everything into a one-page monthly overview with specific items to review.

Human checkpoints. Monthly review of the one-pager; deeper investigation of large variances; review of all flagged transactions.

Variations. Extends naturally to household cash-flow, small-business bookkeeping, or portfolio review. This is the single most useful personal workflow I built, and the template is general enough that a version of it applies to nearly anyone.

———

Closing Note

These templates are intended as starting structures. The value comes from building them, shipping them, discovering where they break, and refining them over months.

Do not overbuild before shipping. A three-agent workflow running in production beats a five-agent workflow stuck in design. Start small. Ship the first week. Refine for the next three.

The goal is not a beautiful architecture diagram. The goal is having agent fleets that quietly handle a growing share of your recurring work, freeing your human attention for the parts only you can do.

Appendix D: *Reading List and Primary Sources*

Every source cited in this book, organized by theme, with a one-line note on why it matters. Where a work is cited in multiple chapters, I have placed it under its primary theme.

This list is deliberately short. Almost every item on it is genuinely worth your time. Skipping the long reading list keeps you from the common trap of collecting books instead of reading them.

— — —

Labor Economics and AI Impact

- David Autor, *Why Are There Still So Many Jobs?* (2015). The pre-generative-AI labor economics argument. Core framework still holds; the predicted edges have accelerated.
- Daron Acemoglu and Pascual Restrepo, collected papers on automation, displacement, and reinstatement. Technical but accessible to a careful reader. Current reality tracks their displacement scenario.
- Hans Moravec, *Mind Children* (1988). Original source for Moravec's Paradox. The single most predictive observation in modern labor economics, published forty years too early to be believed.
- Goldman Sachs, *The Potentially Large Effects of Artificial Intelligence on Economic Growth* (March 2023). The 300 million FTE exposure estimate and methodology. Public, downloadable.

- Anthropic Economic Index, published and updated online. Real-world AI usage data by occupation.
- Stanford AI Index Report (annual). The authoritative benchmark-tracking document. Read the current year's executive summary.
- World Economic Forum, *Future of Jobs Report* (2023, 2025). Skill-half-life and job-category projections; useful directionally.

— — —

Psychology, Mindset, and Adaptability

- Carol Dweck, *Mindset* (2006). Foundational lay treatment of the growth-versus-fixed mindset research.
- Daniel Kahneman, *Thinking, Fast and Slow* (2011). System 1/System 2, intuitive expertise, calibration. Essential reference on judgment.
- Karl Friston's work on the free energy principle. Technical; the layperson's version is captured well in short essays and interviews if the papers are too dense.
- Ellen Langer on mindfulness and aging, particularly the "counterclockwise" research.
- K. Anders Ericsson on deliberate practice. Often misquoted; the actual research is about specific, feedback-driven, uncomfortable practice.
- Viktor Frankl, *Man's Search for Meaning* (1946). Foundation text on meaning-making.

— — —

Creativity and Innovation

- Frans Johansson, *The Medici Effect* (2004). The combinatorial creativity argument.
- David Epstein, *Range: Why Generalists Triumph in a Specialized World* (2019). The modern sequel.
- Margaret Boden, *The Creative Mind* and related work. Philosophical taxonomy of creativity — combinatorial, exploratory, transformational.

— — —

Emotional Intelligence and Presence

- Vivek Murthy, *Our Epidemic of Loneliness and Isolation* (2023 U.S. Surgeon General's Advisory). Free, public, remarkably readable.
- Robert Waldinger and Marc Schulz, *The Good Life* (2023). Synthesis of the Harvard Study of Adult Development.
- Robert Putnam, *Bowling Alone* (2000). The civic-fabric decline argument.
- Carl Rogers on the therapeutic conditions. The core articulation of what presence is in a helping relationship.
- Gary Klein, *Sources of Power* (1998). Naturalistic decision-making and expert pattern recognition.

— — —

AI and the New Literacy

- Ethan Mollick, *Co-Intelligence* (2024) and ongoing writing at oneusefulthing.org. Accessible, current, practical.

- Anthropic's public documentation on tool use, agent architecture, and Constitutional AI.
- NIST AI Risk Management Framework (AI RMF 1.0 and generative AI profile). The U.S. baseline for organizational AI governance.
- MCP (Model Context Protocol) documentation, as it evolves. The plumbing layer for agent interoperability.

— — —

Regulation and Compliance

- The EU AI Act, full text. Read the risk tiers and the obligations carefully if your work touches Europe.
- IAPP (International Association of Privacy Professionals) materials and certification bodies of knowledge.
- Your state bar association or professional licensing board's AI guidance, if you are in a licensed field. Updating quickly; worth checking quarterly.

— — —

Financial Planning and Retirement

- Michael Kitces, kitces.com. Technical planning content, written for planners, accessible to careful readers.
- Bogleheads wiki and forum. Empirical, unglamorous, community-moderated.
- Fidelity, Vanguard, and Morningstar research on retirement savings adequacy. Free, reputable.

- Social Security Administration planner tools. Underused by most Americans.

— — —

Career and Work

- Tim Ferriss, *The 4-Hour Workweek* (2007). Uneven, partially dated; core insights about lifestyle-first business design hold up.
- Derek Sivers, various short books. Unconventional, sharp, clarifying.
- Cal Newport, *Deep Work* (2016) and *Slow Productivity* (2024). Foundational vocabulary for the kind of attention modern knowledge work actually requires.

— — —

For Parents and Young Adults

- Mike Rowe's mikeroweWORKS foundation materials. Trade career advocacy and scholarships.
- Harvard Graduate School of Education, *Pathways to Prosperity*. Research on non-college career pathways.
- Bureau of Labor Statistics, *Occupational Outlook Handbook*. Free, updated, the most useful single resource for thinking concretely about any specific career path.

— — —

How to Use This List

Pick three items. Read them this quarter. Not five. Not ten. Three. The rest wait until you have actually engaged with the first three.

Most of us collect books as a substitute for action. Do not do that with this list. If one book changes how you think about your situation and one article leads you to a concrete decision, the list has done its job.

Appendix E: *Glossary*

Key terms used in this book. Short definitions, alphabetical.

— — —

Accountability Premium. The wage uplift attached to roles whose primary economic value is a licensed human's willingness to accept personal responsibility for a regulated output. The single most durable category of white-collar work in the cortex revolution. Chapter Five.

Agent (AI). A large language model hooked up to tools and given a goal, with enough autonomy to take actions between the moments a human supervises it. Distinct from a chatbot, which only responds to prompts. Chapter Eleven.

AGI (Artificial General Intelligence). Hypothetical AI system with human-level breadth of capability across most cognitive tasks. Terminology contested; used loosely. This book uses it in the loose sense.

ASI (Artificial Super Intelligence). Hypothetical AI system with capabilities substantially exceeding human level across most or all cognitive tasks. More speculative than AGI; included for completeness.

Beta-State Career. The career model in which credentials, skills, identity, and bets are held loosely and continuously updated, as opposed to specialized and locked. Chapter Four.

Beta-State Mindset. The psychological operating system that makes a Beta-State Career sustainable: holding skills as tools, titles as hats, opinions as hypotheses, expertise as scaffolding. Chapter Seven.

Combinatorial Ceiling. The structural limit on AI creativity: AI can only produce combinations represented in its training data, not novel combinations across weakly-connected domains. The space above the ceiling is where the human premium for creative work concentrates. Chapter Eight.

Compliance Native. Designing systems with regulatory requirements built in from day one, rather than bolted on after the fact. Also the career path for professionals who combine regulatory literacy with technical literacy. Chapter Twelve.

Cortex Revolution. The current labor transition, in which AI systems are automating mid-to-high-complexity cognitive work at unprecedented speed. Distinguished from previous Muscle-based industrial revolutions. Chapter Two.

EQ Economy. The set of professions whose primary economic value is genuine human presence, attention, and emotional intelligence. Chapter Nine.

Hybrid Translator. A role category that bridges two previously separate professional domains. Often a highly valuable and under-staffed category. Chapter Nineteen.

Liability Gap. The structural gap between what AI can produce and what a legal system will accept as a binding, accountable professional output. Bridged only by a licensed human. Chapter Five.

LLM (Large Language Model). A neural network trained on vast corpora of text, capable of generating coherent output. The underlying engine of most current generative AI systems.

MCP (Model Context Protocol). A protocol for AI agents to interoperate with external tools and data sources. Becoming the plumbing layer for production agent systems.

Medici Effect. The observation, drawn from Renaissance Florence, that breakthrough innovation concentrates at the intersection of fields rather than at the deep center of any single field. Chapter Eight.

Moravec's Paradox. The observation that what is easy for computers (abstract reasoning) is hard for humans, and what is easy for humans (perception, locomotion, manipulation of the physical world) is extraordinarily hard for computers. The asymmetry defines which jobs survive the cortex revolution. Chapter Three.

New Literacy. The three-level fluency model for AI use: Prompt, Workflow, System. Most knowledge workers are stuck at Level One. Chapter Ten.

Physicality Wall. The structural barrier that protects trade and clinical work from automation, built from unpredictability, dexterity, and economics. Chapter Three.

Presence Premium. The wage uplift attached to roles whose primary economic value is a human's physical presence and genuine attention to another human. Chapter Nine, Chapter Seventeen.

Prompt Architecture. The practice of designing multi-agent AI workflows rather than issuing single prompts. The actual professional skill, as distinct from "prompt engineering" as a dying standalone credential. Chapter Ten.

Signer Role. Any job whose primary economic value is the human willingness and legal ability to accept personal responsibility for a final output. Includes physicians, attorneys, CPAs, licensed engineers, compliance officers, and many others. Chapter Five.

Six-Month Half-Life. The observation that many specific knowledge-work skills now lose half their relevance within six to twelve months, requiring continuous updating rather than one-time credentialing. Chapter Four.

Solo Operator Model. The emerging operating pattern for small businesses in which a single human plus an agent fleet replicates the output of a traditional team. Chapter Thirteen.

Starfish Shape. The recommended career shape for the cortex revolution: five or six arms of medium-depth competence in different fields, plus one deeper central core. The successor to T-shaped and I-shaped career models. Chapter Eight.

Taste Gap. The gap between AI-generated output volume and the human capacity to identify which of that output is actually good. The main economic bottleneck in fields where production has been commoditized. Chapter Six.

Three C's. Contribution, Connection, Craft. The three durable sources of meaning in work, which the twentieth-century career bundled together and which the cortex revolution is forcing many of us to disaggregate and rebuild. Chapter Twenty-One.

Appendix F: *25-Question Human Skills Self-Assessment*

Twenty-five questions across the five dimensions that matter in the cortex revolution: Physicality, Accountability, Taste, Adaptability, and New Literacy. Score each question from 1 (not at all true of me) to 5 (strongly true). Add your scores by category and compute a total.

Take this twice — once now, and again ninety days after finishing the 90-Day Plan in Appendix A. Compare. Changes are the data.

Answer honestly. The assessment is only useful if you are.

— — —

Physicality

Rate each statement from 1 (not at all true) to 5 (strongly true).

103. My work requires me to be physically present in a specific location at specific times.

104. My work requires me to use my hands or body in skilled ways.

105. My work involves navigating unpredictable physical environments or situations.

106. A robot attempting to do my job would face significant real-world complexity beyond what training data can capture.

107. My physical presence adds value that cannot be replicated remotely.

Physicality subtotal: ___ / 25.

— — —

Accountability

Rate each statement from 1 (not at all true) to 5 (strongly true).

108. I hold a professional license that permits or requires my signature on work products.

109. I am personally liable — legally, financially, or professionally — for decisions I make in my role.

110. My role requires credentials that take years to earn.

111. A regulator or professional board can sanction me personally for work failures.

112. My signature or credential appears on documents with legal or contractual consequences.

Accountability subtotal: ___ / 25.

— — —

Taste

Rate each statement from 1 (not at all true) to 5 (strongly true).

113. I regularly make judgments about quality that would be difficult for AI to replicate.

114. I can articulate specifically what distinguishes excellent work from merely adequate in my field.

115. Others in my field seek my opinion on quality calls.

116. I consume high-quality examples in my field deliberately and regularly, not just passively.

117. I can defend my aesthetic or quality judgments in writing or in meetings.

Taste subtotal: ___ / 25.

— — —

Adaptability

Rate each statement from 1 (not at all true) to 5 (strongly true).

118. I learn new tools or methodologies on purpose, not just when required.

119. I have publicly revised an important professional opinion in the last year.

120. I seek out discomfort in my work as a learning signal, rather than avoiding it.

121. I have done something professionally scary in the last quarter.

122. I hold my current job title loosely; it does not define my core identity.

Adaptability subtotal: ___ / 25.

— — —

New Literacy

Rate each statement from 1 (not at all true) to 5 (strongly true).

123. I use AI tools as part of my regular workflow.

124. I have designed at least one multi-step AI workflow for a recurring task.

125. I understand the difference between prompting and orchestration.

126. I can describe how to supervise an AI agent, not just how to use one.

127. I have built something AI-related I could show to a hiring manager.

New Literacy subtotal: ___ / 25.

— — —

Total and Interpretation

Total score: ___ / 125.

100–125: Strong position. You are already operating from multiple pillars of the human premium. Focus on your single weakest subcategory. If New Literacy is below 15, make it your priority for the next quarter regardless of your other strengths.

75–99: Solid foundation with specific gaps. Identify your one or two weakest subcategories. Build toward them deliberately. You have time, and you have a workable starting position.

50–74: Pivot territory. Your current configuration is not durable for the next decade. Use the book's framework to identify a specific pivot target and start the 90-Day Plan in Appendix A. Do not wait.

Below 50: Urgent. Every chapter of this book is relevant to you. Start the 90-Day Plan this week, not this month. The pivot is viable from here, but only if you start soon.

If any single subcategory scores below 10, that is your priority. A total of 85 with a 7 on Accountability is a structurally different situation than a total of 85 with balance across categories. The low pillar is the leak in the boat.

If every subcategory scores above 20, you are in rare company. Keep compounding. The gap between you and most of your peers will widen over the decade, and the value will compound with it.

— — —

A Note on This Assessment

This is a self-assessment, not a psychometric instrument. It will not capture every dimension of career resilience, and it will give you a noisier reading on your first attempt than on your third.

The point is not a precise score. The point is honest reflection across five specific dimensions, run on a schedule, with the results compared across time. The act of taking the assessment quarterly for a year will do more for your career than most external career services you could pay for.

Pair this with the 90-Day Plan. The two instruments work together — the assessment diagnoses, the plan treats.

Appendix G: *Pivot Resources*

Concrete resources for the specific pivot paths discussed in this book. Where possible, I have named credentialing bodies, authoritative sources, and finding mechanisms rather than individual programs — because programs change faster than credentialing standards do.

Before pursuing any pathway, verify current requirements and costs with the specific program or licensing body in your state.

— — —

Healthcare Pathways

- **Nursing (RN, NP).** American Association of Colleges of Nursing (AACN), aacnnursing.org. State boards of nursing for licensing specifics. Your state's community college system for accessible ADN-RN programs.

- **Physician Assistant (PA).** Physician Assistant Education Association (PAEA), paeaonline.org. National Commission on Certification of Physician Assistants (NCCPA) for post-graduate credentialing.

- **Allied Health (Radiologic Tech, MRI, Ultrasound, Respiratory, PT, OT).** American Registry of Radiologic Technologists (ARRT) and analogous bodies for each specialty. Most allied health programs are at community colleges and are among the most accessible healthcare pivots for mid-career professionals.

- **Physician (MD, DO).** Association of American Medical Colleges (AAMC), aamc.org. American Association of

Colleges of Osteopathic Medicine (AACOM), aacom.org. Realistic only for readers with substantial runway.

- **Pharmacy (PharmD).** American Association of Colleges of Pharmacy (AACP), aacp.org.

— — —

Mental Health Pathways

- **Clinical Social Work (LCSW, LICSW).** Council on Social Work Education (CSWE), cswe.org. State licensing boards for specific requirements.
- **Marriage and Family Therapy (LMFT).** American Association for Marriage and Family Therapy (AAMFT), aamft.org.
- **Counseling (LPC, LMHC).** National Board for Certified Counselors (NBCC), nbcc.org. State-specific licensing varies.
- **Clinical Psychology (PhD, PsyD).** American Psychological Association (APA), apa.org. Longer and more competitive pathway.
- **Psychiatric Nurse Practitioner (PMHNP).** Via RN + advanced practice credentialing. Growing rapidly; often the most practical prescribing pathway for career pivoters.

— — —

Coaching Pathways

- **International Coaching Federation (ICF), coachingfederation.org.** The dominant credentialing body. ACC, PCC, MCC levels. Required for serious corporate coaching markets.

- **Center for Credentialing and Education, cce-global.org.** Board Certified Coach (BCC) credential.

- **Specialty Credentials.** National Board for Health and Wellness Coaching (NBHWC) for wellness coaching. Various trauma-informed and evidence-based practice credentials for specific modalities.

— — —

Skilled Trades

- **Apprenticeship.gov.** The U.S. Department of Labor's registered apprenticeship finder. Start here.

- **Your state's apprenticeship office.** Varies by state; search "[state] apprenticeship office" for the official registry.

- **Associated Builders and Contractors (ABC), abc.org.** Non-union trade training.

- **Union apprenticeship programs.** IBEW for electrical, UA for plumbing and pipefitting, Sheet Metal Workers, Ironworkers, Operating Engineers. Each has a local in most metros.

- **Community college trade programs.** Often the most accessible entry point; look for accredited programs with documented placement rates.

- **mikeroweWORKS.org.** Scholarships and advocacy for trade careers. Particularly useful for young adults and families reconsidering college-track assumptions.

— — —

Compliance, Privacy, and Regulatory Pathways

- **International Association of Privacy Professionals (IAPP), iapp.org.** CIPP/US (U.S. privacy), CIPP/E (European privacy), CIPM (privacy management), CIPT (privacy technologist). The dominant certification body in the field.
- **ISACA.** CISA (auditing), CISM (security management), CRISC (risk), CDPSE (data privacy engineering).
- **(ISC)².** CISSP (information security), CCSP (cloud security).
- **Society of Corporate Compliance and Ethics (SCCE), corporatecompliance.org.** CCEP credential and related programs.
- **NIST, nist.gov.** AI Risk Management Framework and related public guidance. Free, authoritative.

— — —

AI Literacy

- **Anthropic documentation, docs.anthropic.com.** Clear practitioner-level content on tool use and agent design.
- **DeepLearning.AI.** Practical courses in generative AI, agent systems, and adjacent topics. Many are free or low-cost.
- **Hugging Face learning resources, huggingface.co/learn.** Open-source-friendly introduction to the technical layer.

- **Ethan Mollick, oneusefulthing.org.** Ongoing practical writing about AI in real work contexts.
- **Build-your-own projects.** The single most valuable form of AI literacy is building something. Pick a small problem and build a three-agent workflow for it. Repeat monthly.

— — —

Financial Planning Credentials and Resources

- **CFP Board, cfp.net.** Certified Financial Planner credential. The dominant professional standard for fiduciary financial planners.
- **NAPFA (National Association of Personal Financial Advisors), napfa.org.** Fee-only fiduciary directory. Use this to find a planner who is not commission-motivated.
- **Garrett Planning Network, garrettplanningnetwork.com.** Fee-only, hourly fiduciary planners for people who need a few hours of help, not a managed account.
- **Kitces.com.** For the reader who wants to understand the field from the inside before hiring anyone. Written for planners; accessible to careful readers.

— — —

Eldercare Pathways

- **Aging Life Care Association, aginglifecare.org.** Professional body for geriatric care managers. Credentialing and professional development.

- **American Geriatrics Society, americangeriatrics.org.** For clinicians in the field.
- **AARP Family Caregiving Resources.** Free, practical, for readers navigating elder care in their own families.
- **Alzheimer's Association, alz.org.** Specific resources for dementia caregiving and training programs.

— — —

How to Use This Appendix

Pick one pathway. One. Not three. Visit the primary credentialing body's website this week. Read the requirements. Identify the realistic on-ramp for your situation. Make one phone call or send one email to an admissions office.

The credentialing bodies are generally helpful. The programs are generally hireable. The mid-career pivot is increasingly common, which means admissions offices are increasingly equipped to help adults like you. You are not the first forty-five-year-old to call and ask.

The first phone call is always the hardest. It is also always cheaper than another year of drift.

Appendix H: *Ken's Personal Playbook*

This appendix is me. Personally. At sixty-two, writing the book you have just read, figuring out my own version of the advice in it alongside you. A lot of career books end with an author who has it all figured out and is here to tell you how to catch up. I am not that author. I am in this with you. Here is what that looks like, specifically, from my side.

— — —

Where I Stand

I am a corporate instructional designer by day. I have three graduate degrees and forty-five years of working life behind me. My wife and I run two small businesses we have built up over the years. I live in Florida with my wife Izzy, a cat named Ziggy, and a garage full of projects that are always about seventy percent finished. I also have 20+ more book projects I have been working on for decades that I will finally have time to complete.

I am, right now, doing most of the things this book recommends. Not all of them. I am still working some of it out. The honest report is that the advice runs easier in a manuscript than in a life, and when I tell you to book the trade school visit this week, I am saying it as someone who took about six weeks to do my own equivalent of it.

Treat this appendix as what a friend would tell you over coffee, not what a book is obligated to say. I am going to be less polished here than in the chapters. That is on purpose.

— — —

My Own Three C's

Contribution. My day job contributes. The books contribute, I hope, in a different register. The work with my wife on her handywoman business — yes, really — contributes in a small and direct way. My volunteer work at a local nonprofit contributes. I notice that on the weeks I do not feel like I am contributing, my mood is measurably worse by Thursday. So I protect contribution like a utility.

Connection. This has been harder than it should be, honestly. I have Izzy, which is most of it. I have a small circle of longtime friends I stay in touch with through a monthly text thread that is ninety percent jokes. I am deliberately trying to build more local connection this year. Moving from Central Florida to Dallas and then back to Florida a while back broke some things, and restoring the fabric has been slower than I expected.

Craft. Writing. Specifically, the drafting hour I try to protect six days a week, early morning, usually before seven. When I skip it for more than three days, I can feel it. Also the garage projects, when I am honest. The old bikes. A motorcycle we are rebuilding in slow motion. The craft arms of my starfish are disproportionately physical, and I think that matters as I age.

I have written this out because writing it out is how I audit it. My own three C's are not in perfect shape. They are better than they were three years ago and they are still a work in progress.

— — —

What I Am Doing This Year

I am doing the Friday Unlearning Hour. Most Fridays. I miss it about once a month, and I notice the miss.

I am building agent workflows for my own business. One to pull beta reader feedback into categorized summaries. One to run distribution reports across six retail channels. One to flag books that are underperforming and suggest what to reconsider. They are not elegant. They work.

I am learning AI voice production — the nine-hour recording session I mentioned in Chapter Seven was real — as a specific application of agent wrangling to my existing publishing work. I was lucky enough to snag a 5090-video card and can do the audio AI work locally now; the results are uneven so far. I am still in the first year.

I am reading outside my lane, deliberately. This year it is behavioral economics, which I have been avoiding for thirty years because I thought it was a marketing trick. It turns out to be much more than that.

I am seeing a credentialed financial planner every twelve months, and I am paying her flat fees rather than commissions because I read Kitces before I hired her.

I am planning to work, in some form, until at least seventy. Yes, because I cannot financially afford to stop due to unexpected life situations. But also because the work is the thing I *want* to be doing, it's fun for me. I also teach ID at work and elsewhere.

— — —

What Scares Me

The pace of change. I am sixty-two. Most of the technology shifts I am writing about are running faster than I am sometimes. There are weeks where the daily progress feels like too much, and I wonder whether the advice I am giving you will be obsolete before the hardcover hits the shelf.

Age bias. I know it exists. I have seen it affect people I respect. I hope this book is good enough that it does some small thing to counter the assumption that people my age are not adapting. I am adapting. Many of my peers are adapting. Not all of us are. I worry about the ones who are not.

The economic transition hitting people I love. I have adult family and close friends whose jobs are in the autopsy chapter, and watching them navigate it is harder than writing about it from a careful distance. Some of them are doing okay. Some are not. I would like all of them to be okay, and I cannot make them so.

Health. I am sixty-two. Health, plural. You know what I mean.

The loneliness point from Chapter Nine. I see it in my community. I feel it sometimes myself, on the longer stretches when work is quiet. I take it seriously.

— — —

What Gives Me Hope

People. Specifically, the readers I have heard from over three years of writing about this transition, who are doing the work, quietly, in small towns and suburbs and cities across the country. The middle-aged nurse who is finishing her doctorate. The insurance professional who re-credentialed into licensed financial planning. The teacher who pivoted into special education because the presence-premium stacks there. The plumber's apprentice at fifty-one. Every one of those people is evidence that the transition is survivable, and that the advice in this book is not theoretical.

Izzy. My wife has built a small business over the last few years that is quietly one of the most agile operations I know. She adapts. She learns the new tools as they arrive. She is not afraid of any of this. Watching her has done more for my faith in the resilience of working people than any data set I could cite.

The structure of the problem itself. The cortex revolution is not the first labor transition. Humans have been through this before, at slower speeds. The advice in this book is mostly restatement of principles that have worked across transitions — adapt, credential, specialize in what stays human, protect your community, build optionality. We know how to do this. We have just not had to do it in our lifetimes.

The book itself, honestly. I have been thinking about this material for three years. Writing it forced me to work out my own plan in detail. I am in a better position now than I was before I started writing, and I suspect the same is true for some meaningful share of readers.

— — —

A Final Word to Readers My Age

If you are over fifty-five and reading this, let me say this directly.

You have more runway than you think. You are worth more than the market is currently telling you. Your judgment, your relationships, your history in your field, are genuinely scarce in a system that has been flattening everybody else's. You are at exactly the age where, with a few deliberate moves, you can convert what looked like liability into leverage.

The specific advice in this book applies to you. The Signer Role. The Presence Premium. The Solo Operator intrapreneur path. Agent wrangling. Compliance native. Every one of these was written with readers your age in mind.

Do the 90-Day Plan. Take the assessment. Make one phone call this week about one credential, or one pivot, or one conversation. Ninety days from now, you will be different. Two years from now, you will be on the other side.

I am rooting for you from my kitchen, with cold coffee, on a Saturday afternoon, like an old friend who happened to write a book about something we are all going through at the same time.

Go get Monday.

— Ken.

— END OF STILL USEFUL in an AI WORLD —

Ken Konet, M.Ed., MBA • Humbolton Press • April 2026

About the Author

Ken Konet has spent twenty-five years as a corporate instructional designer, which is the polite professional way of saying he gets paid to teach working adults things they really do not want to learn but desperately need to know. Before that, he was a kid in a Cleveland, Ohio suburb who read too much, lived in libraries, and has always loved learning. The through-line has been roughly consistent ever since.

He holds an M.Ed. in Instructional Design and two MBAs, which together qualify him to talk about adult learning with one hand and read a balance sheet with the other. Over the course of his career he has built training programs for Fortune 500 organizations across industries that were, at the time, considered bulletproof. Some of them still are. Most of them are in this book.

The job itself is why his bookshelf looks the way it does. An instructional designer becomes a temporary expert in whatever his clients' workforces need to learn this quarter: compliance, underwriting, mental health, manufacturing, patient care, adult literacy, leadership, safety, the economics of the next industrial transition. You absorb a subject deeply enough to teach it, you teach it, and then the client hands you a new domain and you start over. Do that for twenty-five years and you end up with a frankly unreasonable number of things to say, on a frankly unreasonable number of topics. For Ken, most of them will eventually turn into a book.

A lot of those books have been a long time coming. Several of the manuscripts in his catalog, this one included, are projects Ken has been sitting with for two, five, ten… even twenty years in some cases. Notes in spiral notebooks. Drafts on recovered hard drives. Ideas that would not leave him alone. What he did not have, for most of those years, was the time.

Return-to-office changed that. Ken now commutes weekly between Florida and Texas for his day job, which means airplanes, airport lounges, and the strange quiet hours of business travel. As it turns out, the one thing corporate America could not automate out of his schedule was the flight back on Friday night. The manuscripts he had been carrying for a decade finally had somewhere to land.

He lives in Florida with his wife Isabella, who runs a professional handywoman business and is, by unanimous household vote, the most competent person either of them knows. When he is not writing, Ken rides motorcycles, hikes and camps, and plays video games he came to embarrassingly late in life and is only occasionally good at. At sixty-two, he has made his peace with the fact that none of this counts as slowing down.

Still Useful is his field report from twenty-five years of watching professionals learn, unlearn, and pivot. It is also the book he wishes somebody had handed him at forty-five.

Also by Ken Konet

Nonfiction

- *The Happiness Algorithm*
- *Stop Stepping on Rakes*
- *The Discomfort Dividend*
- *The Whole Child*
- *Step One, Alive!*
- *Migraines Demystified*
- *Cervicogenic Migraine: The Missing Diagnosis*
- Nobody Reads Your Resume at Your Funeral
- The Engaged Leader
- Adults Don't Exist
- Move Forward
- The Power of One Word

Fiction

- *PATSY*
- *Ninety Miles Under Radar*
- *Mycelium*
- *Curb Appeal*

For Young Readers

- *The Upsetty Spaghetti series* (21 titles in social-emotional learning for early readers)
- *Happy Halloween*

With Ibrahim Roble

- The Meaning of YOUR Life
- *Lucifera (book 1 of The Lucifera Series)*
- *Mercy is Blind (book 2 of The Lucifera Series)*
- *Dream Heist*
- ***Monsterific**: Midnight's Puppet*
- *Gamify Your Life*
- *Demon Trainer*

The Hundred Acre Wood Monsters holiday series

- Winnie the Pooh and the Vampire
- Winnie the Pooh and the Zombies

- *The Blank*
- *The Heartsmith*
- *The Were-Turkeys*

Also from Humbolton Press

Independent publishing across nonfiction, fiction, and genre work. Selected titles from the Humbolton Press catalog.

Ibrahim Roble

- *The Good Warden*

Paul Green, M.Ed.

- *Bullies Run the World*
- *The Wealth Delusion*
- *Profits of War*
- *Human Again*
- *Stop Being Played*

Jordan Wright, MBA

- *Velvet Order: Book One*
- *The Spider's Web*

Shawn McAllister

- *Friendly Fire*

Heather Hollis

- *Sonnet 155*
- *Aliens vs Gods: Book One*

Kody Foster

- *The Flicker*

www.ingramcontent.com/pod-product-compliance
Lightning Source LLC
LaVergne TN
LVHW031924090826
845145LV00018B/2828

* 9 7 8 1 9 6 6 7 0 3 3 2 7 *